The Wild Berry Book

Romance, Recipes, & Remedies

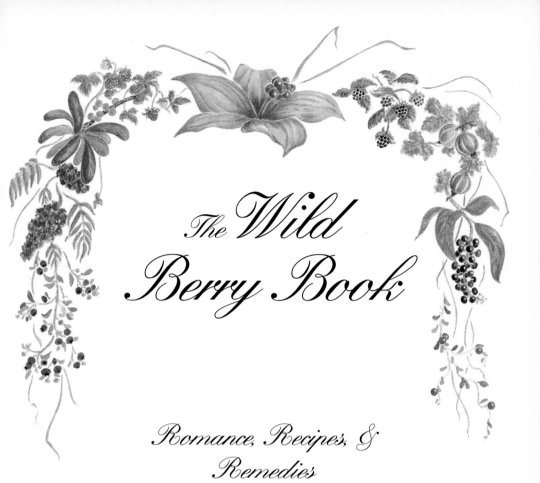

The Wild Berry Book

Romance, Recipes, & Remedies

by

Katie Letcher Lyle

Illustrations by Laurie Anderson Caple

NORTHWORD PRESS

Minnetonka, Minnesota

CREATIVE
PUBLISHING
international

NorthWord Press
5900 Green Oak Drive
Minnetonka, MN 55343
1-800-328-3895

Library of Congress Cataloging-in-Publication Data
Lyle, Katie Letcher, 1938-
    The wild berry book: romance, recipes, and remedies / by
Katie Letcher Lyle.
        p.   cm. — (Camp & cottage series)
    Includes bibliographical references.
    1. Wild plants, Edible.   2. Berries.   3. Cookery (Berries)
4. Berries—Therapeutic use. I. Title. II. Series.
QK98.5.AIL94   1994
582'. 0464–dc20   93-48078

Printed in Hong Kong

Table of Contents

Acknowledgments

Being no botanist, I certainly could not have written this book without the help of many gracious and talented people. I want to thank them here: my editor Greg Linder, for his gentle and wise guidance; my longtime friend and naturalist Janet Lembke, for her help and encouragement; Wylma "Sis" Davis, Margie Camper, and the Virginia Military Institute's Preston Library; Dr. Jim Duke of the United States Department of Agriculture; Dr. Burwell Wingfield, my mushroom-hunting friend and botanical consultant; George Flint of the Village Garden Center in Waynesboro, Virginia; Wendy Knick and Karen Bailey, my horticultured friends; my dear husband Royster Lyle, who continues after 30 years to tolerate and encourage my projects; Dr. Tom Nye and Dr. John Knox of Washington and Lee University, for their generous assistance; Lisa McCown of Washington and Lee University's Special Collections; Jeremy Ledbetter, for his help photographing me and berries; my brother Peter Letcher and sister-in-law Nancy Anderson Letcher for their help; my niece Emily Lyle Kelly for sending me a valuable handbook on wild berries; Grace McCrowell, Chip Barnett, Hester Holland, and the entire talented and good-natured staff of the Rockbridge Regional Library; and the folks at The Blue Ridge Poison Control Center at the University of Virginia, for answering my questions. I am grateful to Neil A. Harriman of the Biology Department of the University of Wisconsin–Oshkosh for his reading and correction of the manuscript.

Photographers

Pat O'Hara: pp. 9, 27, 30, 47, 55, 67, 112; Paul Rezendes/Positive Images: pp. 16, 129 bottom; Lawrence A. Michael: pp. 18-19, 38, 128, 137; Bill Lea/Dembinsky Photo Assoc.: pp. 20, 32; Paul Rezendes: Back cover and pp. 22, 80, 122-123; Gary Milburn/Tom Stack & Assoc.: p. 29; Jerry Howard/Positive Images: p. 35; John Heidecker: pp. 37, 58; Dembinsky Photo Assoc.: p. 40; Larry West: Cover and pp. 43, 124, 132, 143; Mary Nemeth/Photo/Nats: p. 44; John Gerlach: pp. 50, 108, 116; Brett Baunton: pp. 56, 130-131; Frederick D. Atwood: pp. 63, 92, 145; Fox Hollow Enterprises: p. 65; Thomas Boyden: p. 68; Ann Reilly/Photo/Nats: p. 72; Mary Jane Hayes: p. 74; Joy Spurr: p. 76, 96, 110-111, 141; Larry Mishkar: p. 79; Lynn Rogers: pp. 82-83; Dick Scott, Dembinsky Photo Assoc.: p. 84; William H. Mullins: pp. 88-89; Charles Gurche: p. 90; Marilyn Wood/Photo/Nats: pp. 94-95; Gay Bumgarner/Photo/Nats: p. 98; Don Johnston/Photo/Nats: pp. 101, 140; Tom Cawley: p. 102; Kenneth Smiegowski: p. 106; Mark Wilson/WILDSHOT: pp. 114-115; Richard Reddig: p. 118; John and Ann Mahan: pp. 120, 146-147; John J. Smith: pp. 127, 134, 148; Steve Schneider: p. 129 top; Wilf Schurig: p. 135; Greg Vaughn/Pacific Stock: p. 138; William H. Allen, Jr.: p. 142.

Baskets
of Berries

Sweet berries ripen in the wilderness . . .
—Wallace Stevens

Foreword

I feel lucky to have grown up during World War II to the light music of gossiping women, the static, make-do society of the war years, and the perfumy scents all summer long of fruits being put up. When we weren't spending those first free summer days in bored isolation from the dreaded polio, we went running with pails along the railroad tracks for the wild strawberries we knew we'd find. All summer we gathered watercress, or grapes to be simmered into jelly, or blackberries for pies, swapping stories about the enormity of the rattlesnakes that were reputed to hang out in blackberry thickets.

Once home, there was the carol of kettles of water bubbling in tropical humidity for the sterilizing of canning jars, or there were mountains of blackberries to be picked over with purple fingers. On the day we made jelly, pooling our ration tickets for sugar, the children hung around trying to get first crack at the "skim" or scum—that foamy, delicious part that grownups discarded from the top of the jelly and which would prevent its perfect clarity if left on top.

With this book, I welcome the opportunity to recall more of those pleasures, and to introduce them to another generation.

∞

Wild berries may be the only thing of value on this continent found in more abundance today than when the first European settlers stepped onto our soil. Though I had not realized it until I began writing this book, probably not one-tenth of one percent of the wild berries in this country are gathered today.

This book is about wild berries—how to find them and use them, how they got their names, and what part they have played in human history—including how common folk beliefs have been supported or discredited by contemporary research.

Botanically speaking (which I promise not to do too often), a berry is *any fleshy, simple fruit, edible or not, with more than one seed and a skin.* But this includes many fruits not generally recognized as berries, among them grapes, tomatoes, eggplants, bananas, cucumbers, kiwis, quinces, chili peppers, figs, persimmons, pawpaws, and pomegranates. (Berries with rinds, like pumpkins, all citrus fruits, and watermelons, are called *pepos.*)

Red currant flowers

But the *common folk* (non-botanists, including me) are likely to call ovate, many-seeded fruits, *and* other fruits that resemble them visually, *berries*. Strawberries, blackberries, and serviceberries, for instance, are *not* actually berries. Osoberries, or Indian currants (and other currants), are berries. But hackberries, elderberries, and bearberries aren't; they're really drupes, or one-seeded fruits. Serviceberries and chokeberries are pomes, or apple-like fruits.

The folk, ever more practical than learned, have jumbled things nicely by doing what comes naturally, looking with keen observation and inference at plants and coming up with their own system of classification, which goes something like this: "If it looks like a berry, it's a berry." Thus you can begin to see that, between the academics and the rest of us, total agreement on what is or is not a berry is unlikely.

Generally, I will go with the folk opinion, so often more sensible in matters of quotidian life than that of the academic world. However, this book will include some "unlikely berries" such as mayapples. I offer you this handbook on wild berries in the hope that it will bring you many hours of pleasure. But beebleberries, I must warn you, are not found anywhere outside of old Little Lulu comics.

Berry Finder

Often, while walking through the woods or down some country road, you may come upon some berries. But what kind are they? I'll try to give you an idea of what you might have from looking at the ripe berry itself, but don't assume anything; this is only a first step in identification.

BERRY TERMS AND TRAITS: HOW TO LOOK AT PLANTS

First look at the plant itself. Is it an *herb* (like straw-berries), a *shrub* (like blueberries or blackberries), a *vine* (like grapes), a *smallish tree* (like dogwood or juniper), or a *tree* (like mulberry, serviceberry, or hackberry)?

Then look at the arrangement of leaves. Your plant may have the needles of an evergreen, the typical stiff leaves of a palm, or common leaves of various shapes. Leaves can grow opposite each other, alternating on the stem, or spirally around the stem.

Look at the leaves themselves. They may be *simple* (a single leaf blade with a tiny bud at the base), or *compound* (several small leaflets on each stem with a bud at the base).

Leaf edges may be *entire* (which means smooth), *serrate,* or *lobed.*

The berries themselves are important. They may be borne singly, in a *raceme* (on separate stems, or stalks, along an axis), in an *umbel* (growing on separate stalks (stems) from the same level), or in *panicles* (a compound raceme or compound umbel). They may be *terminal* (at the end of the stem) or *axillary* (growing from the angle between stalk and leaf). They may be shiny or *bloomy,* the latter meaning dull or dusty.

A "berry" with one seed, like a plum or cherry, is a *drupe.*

herb

shrub

vine

Q: What is white when it's green, green when it's red, and ripe when it's black? A: A blackberry.

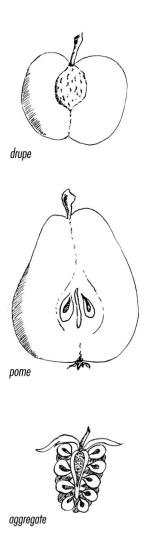

drupe

pome

aggregate

A "berry" that looks like an apple or pear on the inside is called a *pome.* Mulberries and blackberries are made up of a mass of tiny drupes, and are called *aggregates.* Hackberries, rosehips, and some others have the remnants of the *calyx,* or bud covering, showing at the bottom of the ripe berries.

Other terms may occasionally be used. These will be defined as they appear, but the few terms here will help you identify any plant you may come across with fruits on it that look like berries. Look for definitions of health and medical terms in Appendix B.

BERRY NAMES

All plants have genus and species names in Latin. A *genus* is a group of plants that has common structural characteristics distinct from other plants. A *species* is one of a group of structurally similar plants that can interbreed.

In order to distinguish one kind of "deerberry" from another, these scientific names are necessary, supplying a universal language to sort out the confusion that occurs when different folk names arise in different areas, or the same name is given to different plants. Latin notwithstanding, the scientific names often provide information about berries and their histories. *Fragaria virginiana,* for example, is the name of a strawberry first found in Virginia, Virginia's strawberry. *Fragaria* describes an especially fragrant genus.

The Indians had already named plants when the Europeans arrived. In cases where the newcomers had never seen a plant like the one under discussion, they sometimes just accepted the Indian name. (If they observed Indians gather-

ing an unknown berry, they might also have called it an Indian berry.) Indian names suggest (but do not prove) that a plant or animal was indigenous to North America, and was a species the white settlers had not encountered in Europe. Thus we may guess that pawpaws, persimmons, and sumac were berries unknown in Europe. If, on the other hand, the settlers had known something like it back home and already had a name for it, they tended to keep the name they'd always used. Cranberries were known in England, so the settlers kept the name they'd long known; the Algonquins knew them as *ibimi*, which means "bitter berry."

America is a country too big for much consistency, so major confusion can result. Take teaberry, partridgeberry, and pipsissewa; they are three different plants—yet all three are known by the same 10 or 12 folk names, and all are known by each other's names!

Plants resembling other plants are often called after each other, like the mountain cranberries, *Viburnum*, a different but similar-appearing genus than bog cranberries, *Vaccinium*.

Look to the folk names, too, for hints about the berries—if not for perfect truth. Crowberry and curlew berry are folk names of a berry eaten, one supposes, by both crows and curlews (small wading birds). This implies a berry that grows in bogs or near water, and the implication turns out to be accurate. The nickname baneberry should signal a warning to the wise—and the species of *Actaea* are indeed toxic. Bramble berries carry in their name a caution against the briers that protect nearly all of the *Rubus* species.

Sometimes the folk names suggest to us how the berries were used medicinally, like jaundice berry, purgeberry, and goutberry. Some berry folk names

For starters, I offer this exuberantly American berry dessert, which is absolutely failproof. I've never met anyone who didn't love it, and a child of eight could make it. One child I know called it a "berry sexy dessert."

July Church Picnic Berry Dessert (circa 1976)

4 packages ladyfingers
1 cup milk
1 8-oz. package cream cheese
1-1/2 cups milk
1 large package instant vanilla pudding and pie filling
2 8-oz. containers whipped topping, vanilla flavored
1 cup each sliced strawberries, red raspberries, blueberries, and blackberries or black raspberries (or any combination of the above for 4 cups total)
1 cup perfect berries for garnishing—any combination

Line a 9 x 13 baking pan, sprayed with non-stick cooking spray, with the ladyfingers. In a big bowl, first whip the cream cheese with the cup of milk, then gradually add the cup and a half of milk and the pudding mix. Gently fold in one of the whipped toppings. Put half of that mixture in the pan on top of the ladyfingers, then sprinkle the berries over that. Now put the rest of the pudding in. Frost with the other whipped topping, and garnish with the beautiful berries. Let it set for a day before cutting and serving to 12 to 20 people. Note: If it's an Episcopal church, you might sprinkle the ladyfingers with four or five tablespoons of sherry before you get going on the rest of it.

Here's another easy recipe.

Berry Newtons (Jam or Jelly Cookies)

2 sticks real butter
1 8-oz. package real cream cheese
2-1/2 cups all-purpose flour
1 cup any berry jam or jelly

Cut the butter and cream cheese into the flour. When the mixture looks like coarse meal, combine it into a ball and chill it thoroughly for at least two hours. Heat the oven to 350 degrees. Roll out thin, and cut into rectangles three by six inches. (You can reshape and reroll any odd edges.) On one end of each rectangle, put a generous teaspoon of berry jelly or jam. Fold the other half over, press the edges with a fork to seal and decorate, prick the top once with fork tines, and slide with a spatula onto a baking sheet—preferably non-stick. Bake for 10 to 15 minutes, until golden. This makes at least 24 pastries, maybe 36 if you roll them really thin. You can make a real Newton roll, too, by doing a long three-inch-wide roll, folded over the filling. After it cooks, slice it across into "Newtons."

suggest ornamental or practical uses, such as candleberry, necklaceberry, teaberry, and lemonade- or vinegar-berry. Some suggest a berry's habitat, like fenberry or mountain boxberry. Some names suggest botanical characteristics: for instance, bunchberries grow in a tight cluster, and dewberries grow on the ground where the dew forms. And finally, it will not surprise readers that at least a dozen different wild berries are called deerberries, since deer are herbivorous.

What Every Berry Picker Should Know

*N*ot sure if you want to try berry-picking? The reasons why people don't pick wild berries are interesting:

"Exhaust fumes might have poisoned the berries near the road, and I don't dare go after the ones away from the road."

"Don't snakes hang out around berry patches?"

"I stay out of the woods for fear of Lyme disease."

"How do you know when you're getting something poisonous?"

"You mean those red berries are edible? Then how come the birds don't eat them?"

"I can buy them at Kroger's and not risk poison ivy."

"If they're so darn good, how come no one picks them?"

"Ticks!"

"They're not the same blackberries you buy at the store, are they?"

Berry picking is a perfect hobby, in that it costs nothing and yields free food, has few perils, and requires very little in the way of equipment.

Pails with handles are convenient for collecting wild berries, but I prefer large, flat-bottomed baskets, so the berries can be spread out to avoid crushing them. However, even a cardboard box will do nicely.

Though being scratched is something of an occupational hazard for gatherers, berry stains should be regarded as purple badges of courage. Gloves that are thick enough to keep out all brambles are too clumsy and hot to be used, in my opinion. I do recommend long-sleeved shirts and pants, high socks, and sturdy shoes, even boots, to prevent scratches and to deter mosquitoes, biting flies, ticks, annoyed spiders, or mad bees. A spritz

of good insect repellent or a bath in Avon's Skin-so-Soft before you head out helps, too. In areas where chiggers are found, some recommend rubber bands around the leg-bottoms of your long pants, extra bug repellent sprayed around your ankles, or even black pepper in your socks. Bugs hate pepper.

If gnats are a problem, it's helpful to know that they go for white. Wearing a chef's hat would probably be just perfect, but short of that, any white hat perched on your head will keep the gnats out of your eyes. If you don't have a cap, and if the gnats are in your nose, eyes, and mouth, hold your hand above your head. They'll move to your hand.

Snakes sometimes cohabit with berries, but I have it on good authority that snakes possess extremely sensitive ears and absolutely hate whistling. So whistle in the woods. Or sing. Even stomping the ground causes enough vibrations that snakes will get out of your way if given a chance. Hiking boots are a good protection, as snakes rarely strike more than a foot above ground.

Someone once asked me what you should do about bears. The best advice I have on that matter is, if you know bears are nearby, pick your berries somewhere else.

If you are careful, you can avoid poison ivy or poison oak. Learn what each looks like before you go into the woods. If you do get into some, wash it off with soap and water as soon as possible. Jewelweed and plantain are the folk antidotes to poison ivy, so learn those plants also. It's uncanny how often they grow near poison ivy. Just grab some of either, crush it, stems, leaves, and flowers, and rub it over the part of you that the ivy touched. But washing off the oily urushiol, the toxic substance present in the plant resin, is the best way to prevent rashes.

The only real danger I can think of is leaded gasoline exhaust. Berries (or anything else) picked on heavily traveled roads may be tainted—though less so nowadays, I suppose, than before unleaded gas became popular. Pick your berries away from main roads.

I'd suggest wearing old clothes that don't matter, but if you do get stains you can try bleaches and stain removers. However, science has not yet improved much on the following:

The stains of berries, especially of blackberries, and of plums . . . are very refractory. Hence, if possible, these should have immediate treatment. These, when fresh, yield readily to the action of boiling water, especially if the fabric be stretched tightly and the boiling water poured upon it with some force. If stains have been neglected and fixed by soap in the laundry,

it may be necessary to apply dilute oxalic acid or chloride of lime, or to treat them with lemon juice and salt, and other remedies, afterwards exposing the article to the air and sunshine.
—Household Discoveries (1908)

What if you really get lucky? What if you find so much wild fruit you can't possibly use it all at once? I believe the answer is freezing them in a syrup that matches the berry. In each case, boil the sugar and water together until the sugar is absolutely dissolved, then mix it with the fruit, put it in bags or plastic containers, press out excess air, and freeze for future use. Use your own judgment, but I think the following amounts are about right:

Raspberries and blackberries: 6 parts fruit, 1 sugar, 1 water

Gooseberries and currants: 3 parts fruit, 1 sugar, 1 water

Blueberries, huckleberries, and mulberries: 10 parts fruit, 1 sugar, 1 water

Many of the berries are interchangeable and can be substituted in recipes one for another. The only place it may make much difference is in the most popular berry confections—that is, jams and jellies. The reason for the difference lies in the acidity of each berry. Every jam or jelly depends upon accurate percentages of sugar, acid, and pectin, and the balance is delicate. Food scientists tell us that a pH of 2.8 to 3.4, a pectin concentration of 0.5 to 1.0 percent, and a sugar concentration of 60 to 65 percent are required for a good jelling. Failure to jell is the result of too much sugar, too long cooking, or too little pectin. Strawberries, raspberries, and mulberries are all low-pectin fruits, so with them you need to use a powdered or liquid brand of pectin and follow the directions exactly. You can make a jelly eventually without pectin, through long boiling in a wide pan that allows maximum evaporation, but probably at the expense of the fresh berry flavor and always with the threat of scorching.

To determine whether jelly is ready, dip a metal spoon in the hot mixture and hold it above the pan. If two drops move together and form one hanging drop on the edge of the spoon, the jelly is ready to be skimmed and jarred. But following the exact directions in the package of powdered pectin is the best way to assure jelly of good flavor and pleasant consistency.

Berries are like butterflies: They have few enemies. But the deepest reasons for berry-gathering may lie beyond the taste and healthfulness of free, fresh berries. In foraging for any wild food, we may be answering deep signals from the brain, spirit, or body. Who can tell? We may yearn to hunt for sustenance, to trade energy and sweat for nourishment as our ancestors did, and to satisfy one of our most primitive instincts—that of providing food for ourselves and our families.

Meanwhile, the berries wait, ripening in the wilderness.

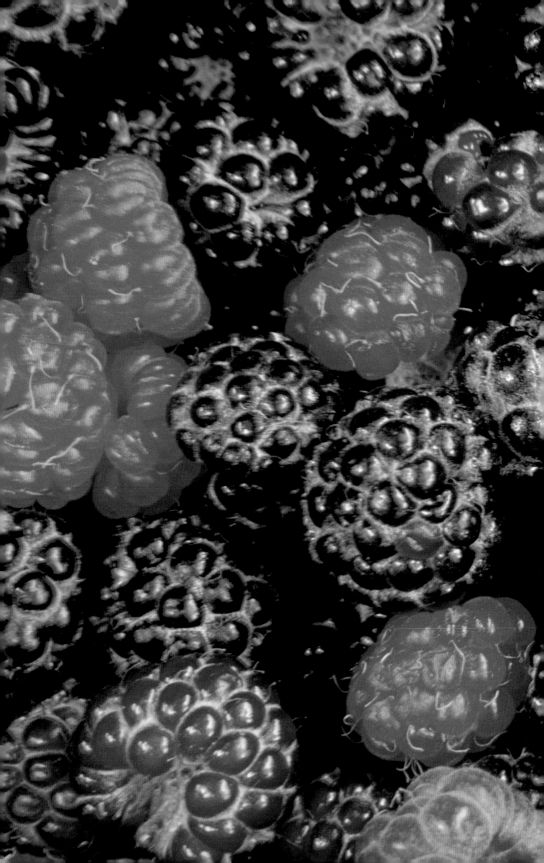

Old Favorites for Eating

*"The strawberye is the wonder of all the Fruites
growing naturally in these partes..."*
— *Sir Walter Raleigh*

Strawberries

*"Strawberries do grow upon hills and valleys,
likewise in woodes and other such places as bee
something shadowie."*
—Thomas Tusser, *Five Hundred Points
of Good Husbandry,* 1597

INTRODUCTION

A strawberry is not a berry, since it bears its one-seeded achenes on the outside of a fleshy receptacle. Nonetheless, it's easily the most popular American berry of all. After all, strawberry ice cream is still America's third favorite kind. (Vanilla and chocolate are still first and second, despite the invention of Ben and Jerry's Cookie Dough flavor.)

Although wild strawberries grow everywhere, they are hard to see. The best patch I ever found was by accident when I was cutting across an open field at the top of a hill in search of a horse. I came back later and gathered a *quart* of wild strawberries, which for me was surely a record amount.

I have heard old-timers tell of making strawberry or peach leather, a chewable confection. The instructions are, as they say, deceptively simple. *Mix equal quantities of crushed fresh fruit and sugar. Spread the mixture out thinly onto sheets of tin, then place the sheets in the sun every day until the mixture has dried to a leathery consistency. When it is done, peel the sheets away, roll them up between pieces of oiled paper, and wrap them in parchment for storage.* My one childhood attempt to make fruit leather was a predictable failure; it should probably have been called Bug Leather, by virtue of the hordes of insects it attracted.

Though the strawberries we mostly find in the stores these days exude only a whiff of the splendor of their smaller wild ancestors, they are still America's favorite berry. Strawberry flavor is so popular that in recent years it has found its way into face creams, shampoos, perfumes, and room deodorants. I once read a recommendation for using crushed strawberries, all by themselves, as a face mask. Put it on, let it dry, rinse it off with warm water. My grandmother told me that as a girl she stained her lips with strawberry juice to make them redder, for her father was "ag'in' paint."

As most know, strawberries, an herb, grow low to the ground beneath their bright green foliage. They have white blossoms and mother plants that put out runners to form new plants. They sport three leaflets on a divided leaf, and the berries (accessory fruits, really) are borne singly and terminally. You will find the ripe fruit a month after the rose-like blossoms appear, though blossom and fruit may be found together. Strawberries may fruit during spring, summer, or autumn; many are now called ever-bearing, which implies that they fruit for a long period.

Wild strawberries are by any economic measure a waste of time, for they are tiny and hard to pick. But they are worth a great deal in superior flavor and pleasure. Gathering these tiny fruits, most of the time no larger than the end of your little finger, is a labor of patient love, as you sit on the ground staining your jeans beyond recovery in an aura of strawberry essence.

Early summer is the time to begin looking for the tiny wild strawberries that were, according to the account of an early settler, so thick in some places that the white feet of cows were stained red as they grazed.

Ideally, they should be gathered in a flat, wide-bottomed basket, so they don't crush each other. But the reward is a flavor so intense it seems fair to say that each tiny wild berry concentrates more flavor than is found in five of the giant modern hybrids. When you get home, red-fingered and sunburned, you would be wise to scrub with coarse soap, rinse, scrub again, and rinse, to exorcise possible chiggers and poison ivy. Then spend a few more hours picking out stems and caps. Never wash wild strawberries. Just pick them carefully.

Wild strawberries seem to grow in the greatest profusion along dirt roads and old railroad tracks. There are two reasons: They like sandy soil, and roads and tracks often have sandy fill; and both locations provide the full, all-day sun that the berries need in order to ripen.

Strawberry Mousse Cake

1 pound cake
1/4 cup kirsch
1 quart clean strawberries
1 cup perfect strawberries
1 cup powdered sugar
1 pint whipping cream with
   a pinch of salt

Cut the cake into one-inch cubes and sprinkle the kirsch over it. Mash berries with the sugar. Let stand for an hour, then stir and sieve. Whip the cream with the salt and fold in the puree. Mix the mousse with the cake cubes and the perfect berries, and chill for six hours before serving. This is a great dessert for a group.

Simply Wonderful Strawberry Shortcake

2 packets or two cups biscuit mix
4 tablespoons sugar
4 tablespoons cooking oil, preferably
   corn or safflower
1/2 cup milk mixed with
1/2 teaspoon vanilla
1 quart fresh ripe strawberries
1 cup whipping cream
a few grains of salt
1 teaspoon vanilla

I make this often. The oil is the secret. Or maybe it's the vanilla. Mix the dry biscuit mix quickly with two tablespoons sugar, oil, milk, and vanilla. Handling as little as possible, pat the mixture out on the bottom of an eight- or nine-inch cake pan that has been buttered and floured. Prick all over with a fork and bake until golden, about 10 to 15 minutes, in a 400-degree oven. Cool. Split it. When you are ready to serve dessert, whip the cream with 2 tablespoons sugar, salt, and vanilla. Withholding one gorgeous berry, spread the berries over the bottom half of the shortbread and let some fall to the sides. Put the top half on, and pour any leftover syrup over it all. Top the whole with the whipped cream, and put the single beautiful berry in the middle. Serves four to eight.

EDIBILITY AND RECIPES

*N*early everyone loves strawberries. The famous American diarist, William Byrd II, ate asparagus and strawberries every day that he could obtain them.

There is one strawberry imitator, the *Duchesnea indica,* or **Indian strawberry**. It is harmless but inedible. People are often fooled by this Indian strawberry or **snakeberry,** which looks like the wild strawberry, leaves, fruit and all, and which has made its way around the world. Its flowers are yellow instead of white, but by the time of fruiting, the flowers have vanished, leaving behind only the clever impostors.

Strawberry shortcake has long been the dessert associated with Independence Day in the U.S., for early July is generally when strawberries are ripe throughout the mid-Atlantic states. Early American chroniclers described "Indian strawberry" bread, a mixture of cornmeal and mashed fruit baked in flat cakes in the coals of a fire.

*T*he word "strawberry" is of unclear origin. The word "streow" meant "hay" and "stray" in Anglo-Saxon; in strawberry plants the runners stray from the mother plant. But *hay* was also often *strewn* beneath the ripening berries to prevent their weight from dragging them down into the mud. Washing much of anything was an idea whose time had not yet come. Furthermore, the berries were in medieval times strung onto *straws* to be sold at market. Some say strawberries have "straws" all over their surface, but that does not prove true on close inspection. Those things dotting the surface are the one-seeded fruits of the plant.

The name of the wild strawberry is *Fragaria virginiana*, apparently suggesting its fragrance and that it was first identified in Virginia. Yet it may be that *Fragaria* comes from an entirely different Latin word, *frango*, I break, from which we derive the word *fragile*. If so, it may relate to the ripe berry's tendency to break easily off its stem. An occasional nickname for strawberries, used in the British Isles, is *hoboys*, believed to be a corruption of *Haut Boys*, a common variety of strawberries. But it may derive from the French *au bois*, meaning "of the woods."

Strawberries, highly esteemed in medieval art, symbolized purity and perfection, and are the plant most often found in conjunction with the Virgin Mary. In France, where they understand the importance of food, it is a custom dating to the French Revolution that the first strawberries of the spring and the first walnuts of the fall must be sent to the President of the Republic. One Madame Tallier, a noblewoman in the court of Napoleon, was reputed to add over 20 pounds of crushed strawberries to her bath water for the health of her skin. Perhaps the berry juice stained it a rosy hue; perhaps some of the generous amount of vitamin C that strawberries contain was absorbed to give her that extra energy and sparkle. Then again, perhaps she just smelled so good that people thought of her as beautiful.

Strawberry Crepes

4 eggs
1 cup mixed milk and water
2 tablespoons melted butter
2 tablespoons brandy
1 cup flour
1/2 teaspoon salt
1 tablespoon sugar
1/2 teaspoon vanilla
1 cup or more of wild strawberries
3/4 cup sugar
1 cup whipping cream
2 tablespoons sugar
1 tablespoon vanilla
1 tablespoon cointreau (optional)

I make these all the time; they're easy and impressive. Blend the first eight ingredients in the blender and let sit for one hour. Brush a six-inch skillet with oil, and pour about two tablespoons of batter in the pan and return to heat, tilting to cover the bottom. When the top appears dry, so that it is cooked on one side, drop it out onto a plate, and cover it with wax paper. Continue until the batter is gone. This makes between 12 and 16 crepes. While the batter is waiting in the blender, mix berries and sugar and let that sit. Whip the cream, sugar, vanilla, and liqueur (if used), and refrigerate. Fold into berries. Put a big tablespoon on the "uncooked" side of each crepe, and roll it up. Garnish with more berries, or sprinkle with powdered sugar. Serves about four.

REPUTED MEDICAL VIRTUES

Strawberries Romanov

2 quarts strawberries
1 cup sugar
1 pint vanilla ice cream
1 cup whipping cream
1/2 cup kirsch, cointreau, vodka,
    brandy, or framboise

Soften the ice cream while you make
an incision in each cleaned and picked
berry. Sprinkle the berries with the
sugar, stir gently, and let sit until all
are "juiced up." Whip the cream,
sprinkle the liqueur on the berries, and
gently fold everything together.
Refreeze for at least an hour before
serving.

*V*irgil (70-19 B.C.), Ovid (43 B.C. to 18 A.D.),
and Pliny (Pliny the Elder, *Pliny's Natural History*,
A.D. 20-76) all mention strawberries. The *Hortus Sani-
tatis*, an ancient Latin text, includes a recipe for a gargle
for throat ulcers: "Take strawberry juice and plantain
water mixed with eight liters of mulberry juice, one liter
of the dung of a white dog, and a little vinegar . . ." The
1597 *Herball* of John Gerard, English surgeon and
botanist, noted that strawberries are "especially good
against all evils of the spleen."

Strawberry juice was valued in the Middle Ages as a
healing potion for sore eyes. Strawberry leaf tea was
believed to have healing properties in general, and to
whiten the teeth. Since ancient times, strawberries have
figured in dissolving kidney stones and curing gout and
gonorrhea. The astringent leaves have been used in a tea
as a diarrhea remedy. The roots, distilled into bitters, are
said to increase urine flow. Carolus Linnaeus, who
named the strawberry, believed that he had cured himself
of gout by eating "quantities of strawberries morning

and evening for several days." In colonial times, strawberries were eaten in great quantity as a cure for consumption (tuberculosis). Some believed that actual cures occurred. Strawberry leaf tea was an Indian remedy for gout, adopted by the American colonists. In addition, a strawberry root tea was used for children with stomachaches, nausea, or bowel problems. Because of the acid in the berries, they are a natural dentifrice, perhaps even dissolving tartar.

There appears to be only one naysayer: the 12th-century herbalist and divine, Hildegarde of Bingen. She believed that strawberries caused illness, pointing out that in June, when they ripened, earaches, tonsil infections, sore throats, rashes, and appendicitis increased dramatically. Looking for an explanation, she hit upon strawberries. Pondering her prejudice, it does seem that an unusual number of people have allergic reactions to strawberries. Hildegarde, in many regards ahead of her time, might have observed this and concluded that strawberries were, at least some of the time, poisonous.

RECENT UPDATE

*T*he huge berries in American supermarkets are crosses between the American wild strawberry and a South American wild strawberry. However, the Driscoll company in California is working hard to overcome bad press about the flavor of commercial strawberries; the company raises 50,000 new cultivars each year, searching for designer varieties that meet market requirements—the first of which, they stress, is *good taste.*

Red Red Strawberry Jam

1 quart wild strawberries, washed and hulled, or good home-grown ripe strawberries
4 cups sugar
juice of one lemon, strained

This jam is beautiful and delicious, but it does not jell. It's great if you don't mind runny preserves. Add sugar to berries and stir gently to keep the berries as whole as possible. When the mix is juicy, cook it slowly, raising the heat gradually to medium. Stir often but gently. When the pot begins to boil at the edges, set a timer for 20 minutes and avoid further stirring; instead, regulate the heat from the bottom so the pot bubbles vigorously but does not boil over. When the timer goes off, carefully remove the pot to a trivet, sprinkle the top with the lemon juice, and allow to cool. When just warm, stir gently and ladle into six or seven one-cup sterilized jars.

Blackberries

And the running blackberry would
adorn the parlors of heaven
—Whitman, Song of Myself: 31

INTRODUCTION

I recall during World War II the women and children pooling gas ration stamps for a trip to the mountains one early summer day, returning with pots, cans, and cooking vessels brimming with black-caps, and everyone with stained scratches that stayed purple for a long time. My mild-mannered mother killed a copperhead with a big stick; the other women came around to praise and thank her, and advised her how to get the spatters of snake blood out of her shirt. That incident stands out as an idyll, an instant of women cooperating, of children attentive and helpful. We were holding the world together. After the fathers came home from war, the women never killed snakes anymore.

In Virginia, we have a legend that the biggest rattlesnakes hang out in blackberry patches, but herpetologists tell me that the scaly brothers hang out wherever they can catch rodents or lizards for dinner—the berries being irrelevant. All of which says you might find a big rattler in a berry patch, but if you did, he'd be there by accident. I suspect it's a story concocted some time ago to keep people out of the concocter's black-berry patch.

DESCRIPTION AND ETYMOLOGY

B rambles grow on every continent on earth. They may be one of mankind's oldest foods, for their seeds are found regularly at Neanderthal sites.

Blackberries are perennial, mostly prickly, shrubs that can be found

along roads, in fields, crowding into the most persnickety gardens, and on the bright edges of shadowy forests and farm ponds during mid to late summer. Some thrive in the shade, though most like full sun. Here are a few common ones:

The **common blackberry** or **highbush blackberry**, classified as *Rubus allegheniensis*, produces white flowers in elongated racemes blooming on second-year stems two months or so before the berries begin to ripen. Blackberries normally display groups of five leaflets growing on a shrubby bush with stalks *either* erect or high-arching, the leaves a grayish green. These grow throughout the eastern states, south to Georgia, north into Canada.

17th-Century Blackberry Wine

Take one Bushell of Blackburys, nine quarts of water, six pounds of sugar; boyle your water and sugar together; then bruize your blackburys in a marble mortar; when they be cold put them together and let them stand 24 hours; then put it into a vessell that will fill and let it stand one month, and then if it be clear bottle it off.

EDIBILITY AND RECIPES

Fortunately, all blackberry and raspberry relatives are wholesome, so if you like the taste of any wild *Rubus*, climbing or resting, whatever color, solid or hollow, gather it. The flavors of all the *Rubi* are characteristic and similar. That the flavors are widely favored is proven by the popularity of jellies and jams, ice cream and pies, and even by fake fruit drinks for children and cosmetics for grown women. The relative sweetness or sourness of the berry will dictate how much sugar to add, but all can be made into anything you'd want to make out of blackberries. Raspberries and blackberries are low in calories (around 50 per cup), high in fiber, and relatively high in potassium.

HISTORY AND FOLKLORE

Throughout history, blackberries have not only fed people but tinted their garments, soothed their illnesses, and even served them as ink. People have long appreciated the high vitamin C content of blackberries and their kin, even if they didn't know what it was. They

Blackberry Wine (#2)

Squeeze berries in cheesecloth to extract juice and pulp. Add an equal amount of water to the juice, pouring it over the seeds and skins. Stir daily, let set a week, and strain. Add 1 pound sugar to each gallon of liquid. Put in bottles loosely topped and leave them until fermentation stops, which will take three to six weeks. To tell when the bottles are ready to cork, listen. If you can hear bubbles, don't tighten or stopper the bottles. Wait until they are silent.

Blackberry Wine (#3)

In Appalachia, the "receipt" for blackberry wine, a traditional Christmas treat for both the recipient and the maker, is simple: Mix a gallon of ripe blackberries in season with two pounds of sugar and a quart of boiling water. Stir the ingredients together, leave them overnight, put through cheesecloth, and bottle. The bottles will, according to mountain dwellers, be ready by Christmas.

probably noticed that people who ate fresh and dried berries didn't often get sick, and that was all the evidence they needed. The berries of *Rubus* varieties, as well as blueberries, were used to dye the popular mauve ribbons so beloved of girls who lived in the 17th century. Blackberries (and blueberries) were sold to Navy-blue merchants into the 19th century for the dying of sailors' uniforms. A dark gray dye has been made by boiling the branches and roots. Blackberry juice has been used in this century to stamp the appropriate grades on meats inspected by the U.S. Department of Agriculture.

REPUTED MEDICAL VIRTUES

*R*ubus berries hold an honorable and consistent place in folk medicine. Roman use dates from around 600 A.D., when Apuleius published his herbal, in which can be found the drawing of a blackberry, *Rubus fruticosus*. Blackberry was believed by Herodotus to cure gout. One of its folk monikers is **goutberry**, perhaps because eating the high-fiber, low-calorie blackberries may ease a digestive system too long marinated in the rich, fatty foods that caused gout (itself derived from the French word for *taste*) in the first place. However, blackberries are not known to inhibit purine formation in the joints, which causes gout.

Teas made from *Rubus* leaves or roots are astringent, and have been used by many primitive peoples. John Gerard's 1597 *English Herball* lists both leaf and fruit as astringent.

In the U.S., the **salmonberry**, a light, rosy-colored blackberry, was used by the Indians of the Pacific northwest as a diuretic in the form of a tea made from the canes. It is said to taste better cooked than raw.

In early American medicine, blackberry leaves were believed to counter diabetes, and blackberry root extract was the major ingredient in a 17th-century Virginia cholera cure. As the leaves are astringent, tea made from the leaves has been used for ages to staunch things that

Blackberry Jam Cake

1/2 cupful butter
2/3 cupful sugar
1 cupful flour
2/3 cupful stoned [seedless] raisins
1 cupful blackberry jam
2 tablespoons soured cream or milk
1/2 teaspoonful soda
1 small nutmeg, grated
1/2 teaspoonful ginger
2 eggs

(This is taken from Mrs. Governor John Letcher's receipt book, written in her own hand circa 1870.) Beat the butter to a cream, then beat in the sugar. When very light, beat the jam in, then the nutmeg. Dissolve the soda in one tablespoon of cold water and add it to the sour cream. Add this and the egg well-beaten to the other ingredients. Now add the flour and beat for half a minute. Sprinkle a tablespoonful of flour over the raisins and stir them in. Pour the batter into a well-buttered loaf pan and bake for 50 minutes. The yield is one small loaf, almost as good as plum pudding.

need contracting, closing up, shrinking, or stopping. These include diarrhea, dysentery, uterine hemorrhage, and sometimes even runny noses.

Rappahannock Indians in Virginia used the roots and berries in steeped tea to allay diarrhea. Cruso's 1771 *Treasury of Easy Medicine* advises soaking blackberry leaves in hot wine and placing them on ulcers morning and evening. The fruits have been eaten to cool fevers. Many of the early settlers believed that the *Rubus* species had special efficacy against "summer complaint," a diarrhea brought on, it was believed, by heat, but more probably caused by spoiled food in an era before refrigeration. On August 4, 1804, one James Anderson wrote a testimonial that was printed in the *Philadelphia Medical and Surgical Journal* the following year:

Last summer, when I was near the settlement of the Oneida-Indians (in the state of New York), the dysentery prevailed much, and carried off some of

the white inhabitants, who applied to the Indians for a remedy. They directed them to drink a decoction of the roots of Blackberry-bushes which they did, after which not one died. All who used it agreed, that it is a safe, sure, and speedy cure.

It has been reported that blackberry leaf juice will fasten loose teeth back into the jaw. Perhaps this is because the juice seems to "pull" at the flesh of the mouth, to "tighten" it. Some Indian tribes used blackberry root steeped in water

as an eyewash; others favored an eyewash made of a tincture of red raspberry leaves. (Tincture means an alcohol solution, so I doubt if that is correct; still, I am loathe to change folk wording. My impression is that tincture can also mean a weak mixture of something.)

Blackberries are said to cure anemia, according to the Doctrine of Signatures, the belief that the color and/or shape of anything in nature dictates its use to man. Thus yellow plants are useful for kidney problems and walnuts are beneficial for the brain. By this reasoning, the black-red blackberry juice might darken "pale" blood. A 19th-century commercial gargle for sore throats, known as Blackberry Glycerite, combined glycerin and blackberry vinegar. A medicine named Blackberry Balsam was marketed in this country as a diarrhea remedy from 1846 until well into the 20th century. The dried bark of the rhizome and roots of *Rubus* plants appeared in the official United States Pharmaceutical list from 1820 until 1916 as an astringent and tonic.

The berries are a popular seasoning for medicines to this day. One contemporary midwestern writer described watching her mother in this century dry fresh blackberries in a low oven and grind them to a powder, to be used as a diarrhea remedy. The powder was to be taken, she wrote, a teaspoon at a time, stirred into hot water. *Rubus* leaves have been analyzed and found to contain a number of medically useful compounds.

RECENT UPDATE

*M*ost recently, blackberries have been the subject of much cross-breeding, in modern science's endless search for the perfect berry: one that will resist mold, ripen slowly, "hold up" in transport, grow big, be tender and juicy, and taste like ambrosia.

*We can any afternoon discover a new fruit
there, which will surprise us by its beauty or
sweetness. So long as I saw in my walks one
or two kinds of berries whose names I did not
know, the proportion of the unknown seemed
indefinitely, if not infinitely, great.*
—Henry David Thoreau

DESCRIPTION

Raspberries (*Rubus idaeus, R. occidentalis*) were named, according to naturalist Euell Gibbons, after an already existent French wine called *raspis.* Wild red raspberries, *Rubus strigosus,* are native to Britain and naturalized in North America. They grow on upright stems with bristles, producing three to five leaflets with whitish down underneath. The fruit, of course, is light red, bloomy, and of excellent flavor. Often two crops a year are produced—one in early summer and one in early fall. Red raspberries have fewer seeds and are perhaps juicier than black raspberries. Raspberries are unusual in that they mold within two or three days after being picked ripe (and of course are not good if they're picked green), which makes marketing them extremely expensive.

Raspberries differ from blackberries in that the mass of drupes, which make up the fruit, are separate from the receptacle and attached only to each other. This translates to hollow berries. *Rubus idaeus* and *R. strigosus* are both **red raspberries.**

Red Raspberries

EDIBILITY AND RECIPES

All red raspberries are prized. Try this simple raspberry cream, at right, from Amelia Simmons' first American cookbook, recorded just as it was printed in 1796.

ETYMOLOGY, HISTORY, AND FOLKLORE

Rubus idaeus was named for Mount Ida on Crete, which was a renowned playground for immoral, immortal lovers. Raspberries were so rare in the 19th century that when Martin Van Buren campaigned for president in 1840, he was accused of "wallowing in raspberries"; that is, practicing extreme profligacy in his campaign.

REPUTED MEDICAL VIRTUES

The red raspberry, along with the strawberry, was used medicinally by ancient Greeks. It was believed that the berries whitened teeth and benefited those suffering from gout. Today, many varieties of *Rubus* are used in wines, cordials, vinegars, and brandies—for uses medicinal and otherwise.

The moistened leaves of *Rubus* plants have been used as poultices to draw bad blood from wounds and festering sores. And some Indian tribes found the soft leaves of some *Rubus* species useful as toilet paper.

A tea made by crushing the dried root in the fall is considered antiscorbutic, a tonic for the uterus, and a

Raspberry Cream

Take a quart of thick sweet cream and boil it two or three wallops, then take it off the fire and strain some juices of raspberries into it to your taste, stir it a good while before you put your juice in, that it may be almost cold, when you pot it to it, and afterwards stir it one way for almost a quarter of an hour; then sweeten it to your taste and when it is cold you may send it up. (A wallop is the time it takes for a boil to rise to the top of the pan. You then remove the pan from the heat source and let it calm down, then repeat.)

febrifuge. One tablespoon of the crushed dried root is added to a pint of boiling water. When cool, it is ready to use, though folk sources are not in agreement upon how much should be taken and how often.

RECENT UPDATE

A substance in the leaves of red raspberries variously called *framamine, fragrine,* or *fragerine,* isolated in the 1940s, strengthens, relaxes, and tones the uterus, thus exonerating the old wives' advice that raspberry leaf tea should be taken throughout pregnancy, labor, and postparturition. The leaves, according to modern analysis, are also high in magnesium, which is still in use to prevent miscarriage.

Berry Vinegars

Pick over two pints of the fruit you are going to use to flavor the vinegar: raspberries, blackberries, strawberries, elderberries, gooseberries, etc. Place them in a crockery jug or jar, and pour two quarts of good cider vinegar over them. After a day, pour the whole thing into a cheesecloth bag and let it drip until no more juice comes out. Save the liquid, throw away the fruit pulp, clean the earthen jug or jar thoroughly with hot water and soap, and repeat the process, using two more pints of the same fruit. Do this three or four times over three or four days for a strong flavored vinegar, being careful to wash the crock thoroughly between each day's new fruit.

Iced Raspberry Vinegar

Mix water, sugar, and raspberry vinegar to taste, and serve over ice for a refreshing summer ade.

Raspberry Charlotte

Butter a pudding dish and cover the bottom with dry breadcrumbs. Put a layer of ripe raspberries over it, sprinkle with sugar, then add another layer of crumbs; proceed in this way until the dish is full, with the last layer consisting of crumbs. Put bits of butter over the top and bake, with a plate over it, for half an hour. Remove the plate and let it brown just before serving. Use only half the quantity of crumbs that you do of the fruit. Eat with cream. Taken from *The Book of Forty Puddings* (1882).

Black Raspberries

DESCRIPTION

*B*lack raspberries, *Rubus occidentalis*, create thickets of wickets as they grow, bending gracefully from their own weight, and putting down new roots where they touch the ground. In winter, this habit of the black raspberry is recognizable as a tangle of arches made of thorny canes, or stems, which grow to be six to eight feet long. The plants spread quickly, creating huge, nearly impenetrable patches of black-caps that make picking the ripe berries a fairly treacherous undertaking. The brier patch that Br'er Rabbit begged not to be thrown into was almost certainly a black raspberry patch. They are native, and they have hooked prickles should you care to look. First-year black raspberry stalks don't produce fruit, and are greenish-blue. The second-year stalks are reddish and waxy. The berries are dark purple instead of red, and seedier than red raspberries.

If you are not inclined to brave the tangles of wickety thorns, you can buy black raspberries from roadside stands or at farmers' markets. If you're intent on taming them, you might plant several small patches in a field instead of one big one, so you can pick around the edges.

EDIBILITY AND RECIPES

*B*y many accounts, black raspberries are the best-flavored of all berries. They are wonderfully aromatic, delicious on cereal, over ice cream, or in pies, and they make a dark, perfumy jelly.

Black Raspberry Mousse

Syrup made of 1 cup water and
1 cup sugar, boiled for five minutes
1 quart black raspberries, cleaned and picked over
juice of one lemon
1 cup perfect raspberries
1 envelope gelatin softened in
1/2 cup water
4 eggwhites, beaten stiff
1 cup whipping cream, whipped with
1 teaspoon vanilla and
1 tablespoon sugar.

Add raspberries to syrup and cook for just a few minutes until soft. Strain through cheesecloth. Add lemon juice. Add gelatin, and stir thoroughly. Cool, fold in eggwhites, then fold in whipped cream. Chill in a pretty mold or a soufflé dish. Garnish with whole raspberries.

Thimbleberries

Katie's Homemade Berry Jellies

In making jellies, I am an advocate of the pectin method, because the less cooking you do, the closer you get to the natural flavor. The powdered pectins include raspberry, blueberry, and blackberry recipes, but less frequently one for black raspberries. Follow a raspberry method if none is available for black-caps or wineberries. Follow the instructions precisely.

Flummery

1/4 cup cornstarch
1 cup sugar
1/4 cup water
a few grains of salt
2 cups juice extracted from wild berries

My father used to make this fruit pudding with wineberries, black raspberries, or blackberries. Mix the first four ingredients together and stir until smooth. Gradually add juice. Heat to boiling and boil gently for one minute, stirring constantly until the mixture thickens and clarifies. Remove from heat. Pour into four individual serving dishes and chill thoroughly. Serve with whipped, sweetened vanilla-scented cream.

Black Raspberry Ice Cream

1 quart ripe black raspberries
1 cup sugar
1 quart whipping cream
1/4 teaspoon salt

This ice cream tastes much better than those made with more cholesterol-wise substitutes. Mash the berries and sugar together. If you object to the seeds, squeeze the sweet juice through cheesecloth. Mix with the cream, and churn-freeze. If you must still-freeze it, remove and beat until smooth as it begins to harden at the edges. Repeat this operation three times to ensure smoothness before allowing it to freeze completely.

ETYMOLOGY, HISTORY, AND FOLKLORE

*B*lack raspberries, called also **black-caps** and **thimbleberries** (possibly because they are hollow, like caps or thimbles, and would fit over the fingertip of a child) ripen one month earlier in any given area than blackberries.

When pioneer John Lewis, with his sons William, Thomas, Andrew, and Charles, explored the wilderness at the western range of the Valley of Virginia in the mid-18th century, they noted the profusion of wild black raspberries. A western variety of the **thimbleberry** (*R. parviflorus*) has a soft, velvety, juicy texture, is bright red, and is common throughout the Pacific northwest—as black raspberries are common throughout the mid-Atlantic states.

The French monks who make Chambord, the rich, black-raspberry cordial, and Framboise, the dry, fruity *eau-de-vie*, reportedly distill 20 to 40 pounds of berries for each liter of their precious liqueur.

REPUTED MEDICAL VIRTUES

*O*jibwas used black-raspberry root tea as well as the berries for stomachache. They were used interchangeably with red raspberries.

Rubus Kin

INTRODUCTION

*M*inor subspecies of the genus *Rubus* abound in different sections of the country; blackberries and raspberries interbreed freely, and can only be generally identified. One subspecies, the **strawberry raspberry**, also called **balloonberry**, is said to taste like strawberry.

The common blackberry has over 200 brothers, and most of the berries ripen at roughly the height of summer plus one month in any given area. (However, there is an early-ripening southern *Rubus, R. betulifolius*, called for obvious reasons **mayberry**, and the **southern dewberry**, *Rubus trivialis*, which fruits in early June.) The *Rubi* that grow on trailing vines on the ground are often called **dewberries**. *Rubus villosus* (or soft-haired) is their name, possibly because the canes trail like hair. *R. canadensis* is a smooth-caned wild blackberry, also known as dewberry and first identified in Canada. *Rubus procerus*, the **Himalaya berry**, has a long berry and was first introduced from Asia through Europe as a garden variety. Having escaped its formal beginnings, it slums about mainly in the mid-Atlantic states.

In swampy areas of the northwest and northeast all the way to the Arctic circle, one may find *Rubus chamaemorus*, an exotic reddish-golden, ground-trailing, knotted-stem blackberry (giving rise to one of its many folk names, **knotberry**) that ripens in late summer and goes by other folk names including **cloudberry** and **baked apple berry**. The names suggest its big aggregations of drupes, like scoops of cumulus cloud, and its subtle, spicy taste, though one source says it is called cloudberry because it is found high in the mountains, up near the clouds. Its epithet, *chamaemorus*, means "ground mulberry," and its fruit does resemble the white mulberry's.

Cloudberry One-Dish Dessert

1 cup sugar
1 cup self-rising flour
1/2 cup milk
2-1/2 cups of cloudberries (or other)
A syrup consisting of 2 cups boiling
    water, 1 cup dark brown sugar,
    and 2 tablespoons of butter.

This is a fun treat for children to make. Heat the oven to 400 degrees. Mix the sugar, flour, and milk in a casserole. Stir in the berries. Pour over the syrup. Bake 25 to 30 minutes, until golden and bubbly.

Knotberry Liqueur

a quart or so of berries
a quart or so of sugar
1 cup water
l liter of vodka

Stir together and let it macerate for at least a month before straining through triple cheesecloth and corking.

Cloudberry

There is even a *Rubus* with blonde drupes called, naturally if contradictorily, the **white blackberry**.

A northwest species, known as *Rubus spectabilis*, (the spectacular *Rubus*) or *R. parviflorus* (the little-flowered blackberry), is an eight-foot-tall, bushing bramble with flowers that tend toward light purple or pink and fruits that are the light, rosy red color of salmonflesh. Its folk name, not surprisingly, is **salmonberry** (in the 19th century, it was sometimes called **roeberry**), and it is found from Canada to California and as far east as Minnesota.

WINEBERRIES

*W*ineberries (*Rubus phoenicolasius*, which means "brambles with purple hair") grow wild in at least the northeastern quarter of the continental United States, having escaped from cultivation. The shiny, slightly waxy or sticky berries of brightest red, with three serrated leaflets, grow on vigorous, arching canes. The fruit falls intact from a dry center receptacle, leaving the berry hollow. Wineberries fruit after black raspberries, but before blackberries.

Wineberries are a favorite berry in the part of Virginia where I live. They resemble raspberries, but the flavor is much tarter. Whereas raspberries are dry with a bloomy finish, wineberries are shiny with a sticky or tacky surface. Euell Gibbons, one of the few naturalists treating the wineberry, claims that it is "a recent immigrant from Asia," but that is not what I grew up hearing. They have now spread to all the middle Atlantic states, but according to local legend, they were introduced around the turn of the 19th century, east of the Blue Ridge on his estate near Charlottesville, by none other than that greatest of all American horticulturists, Thomas Jefferson.

Today we find locally a berry that is neither raspberry nor blackberry, but is certainly kin to

A Simple Dessert

1 quart fresh cleaned blackberries, wineberries, red raspberries, or dewberries
1 cup heavy cream
a few grains of salt
1 teaspoon real vanilla
1/4 to 1/2 cup sugar, depending on the sweetness of your berries

There's nothing better in the universe, as far as I'm concerned. Whip the cream with the salt, vanilla, and sugar until stiff. Fold the berries in gently, and serve immediately. Don't make this ahead of time.

Shrub

1 gallon wild berries
1 quart cider vinegar
2 cups sugar
2 sticks cinnamon
20 cloves
This is a summer drink of yore, and it's usually made from blackberries or raspberries. Let the flavors marry for 24 hours, then bring the mixture to a gentle boil, cool, and bottle. To serve, add water and ice to taste.

Danish Berry Pudding

1 heaping cup each of blackberries, red
    raspberries or wineberries, and red
    currants
1 cup sugar
1 cup white wine
3 tablespoons cornstarch
3/4 cup purple grape juice
1 cup sour cream

Reserve a few perfect berries for
garnish. Stir the cornstarch into the
wine until it is smooth. Mix fruit, wine,
cornstarch, and grape juice thoroughly,
then add 1/2 cup of sugar. Stirring
constantly, bring the mixture to a
gentle boil and boil it until the cloudi-
ness clarifies. Taste, and add more
sugar if you prefer. Chill the pudding
thoroughly and serve with a dollop of
sour cream on top of each serving.
Serves six to eight.

Raspberry (or Currant or Strawberry) Syrup

1 gallon wild berries
1/2 gallon sugar
1 cup water
Mash together and let stand overnight.
Bring slowly to a boil, then boil for 30
minutes. Force through cheesecloth and
let cool to room temperature. This
syrup can be used to soothe sore
throats, pour over ice cream, waffles,
or pancakes, or even as a base for ice
cream.

both. Wineberries, though delicious fruit, make miser-
able wine. They make a clear red jelly that is tart and
delicate, similar to currant jelly, and they make wonderful desserts with whipped cream or ice cream.

RECENT UPDATE

Blackberries and raspberries are crossbred today
by horticulturists in search of the Ultimate
Berry; thus, many new subspecies of *Rubus* have been
developed in recent years, and many of these escape back
into the wild. The **loganberry**, for instance, seems to be
a warm-climate, blackberry–red raspberry hybrid or, as
one writer put it, "the red-fruited sport of the wild
blackberry." It was named for its discoverer, one Judge
Logan, who found the berry growing in his California
garden in 1881. It now grows wild in Europe and is
popular there.

Boysenberries, a more recent miracle, were first cultivated by Rudolph Boysen in California during the
1930s. They have been described as "a wonderful hybrid
of blackberry and raspberry, large and juicy without a
hard core." Another cross, **nectarberries**, are said to be
delicious and winey.

Youngberries, developed by American horticulturist
B.M Young, are a cross between blackberries and a
species of dewberry. The **bababerry** is a southern-
growing red raspberry that was commercially introduced
to compensate for the fact that red raspberries do not
thrive in heat. **Ollalies** are a commercial species grown
on the West Coast and are a cross between a loganberry
and a youngberry. **Tayberries**, a splendid cross first
grown along Scotland's Tay River in 1977, are reputed to
be almost as delicious as wild berries. And so we come
full circle.

Blueberries
& Huckleberries

Peaches are unquestionably a very beautiful and palatable fruit, but the gathering of them for the market is not nearly so interesting to the imaginations of men as the gathering of huckleberries for your own use.

—Henry David Thoreau

INTRODUCTION

We used to go every August when I was young, all six of us, to gather huckleberries along the sandy edges of certain Blue Ridge mountain roads. I can still hear them falling into the coffee cans we carried to collect them. Eventually, I would become aware of silence as everybody's bucket bottoms were covered and no more "pings" could be heard. That evening, my mother would make a dense yellow cake with huckleberries in a loaf pan, always flouring the berries before she folded them in, claiming that the flour kept them from all sinking to the bottom of the batter. The cake was served hot out of the oven with a hard sauce—a heart-melting (though possibly artery-hardening) mixture of butter, heavy cream, powdered sugar, and liquors, beaten to dreamy peaks with a hand mixer, the top of the mound dusted with nutmeg. The cold sauce melted into the hot cake, and it was hard to conceive of anything in the world tasting better. We did this but once a year, because in the Virginia mountains huckleberries (or blueberries) were scarce. I think they still are, and I no longer know any good picking spots. An interstate has cut through the area where we used to hunt.

Although they are similar, it seems to me there are three differences between huckleberries and blueberries: Huckleberries are darker, shinier, and purpler than the lighter, bloomier blueberries; huckleberries are smaller than blueberries; and huckleberries have a more concentrated flavor than blueberries. Further, I think of huckleberries as southern, whereas blueberries are more inclined to cooler climates. Br'er Rabbit stole *huckleberry* jam that belonged to Br'er Fox; *blueberries* were used to dye the navy blue garments of New England whalers. Huckleberry Finn, surely one of America's most-loved literary heroes, was both wild and southern.

DESCRIPTION AND LOCALITY

*B*lueberries and huckleberries are declared by some experts to be the same thing. Euell Gibbons says that, in the northeast, they tend to be called blueberries; in the south, huckleberries. But technically, huckleberries belong to the *Gaylussacia* species, closely related to the *Vaccinia*. In any case, finding a difference with the naked eye can be impossible. Huckleberries are deciduous shrubs with alternate, simple, large or small leaves dotted with tiny resin globules. Blueberries are shrubs with alternate, simple leaves, large or small. Preferring (with Euell Gibbons) to "eat rather than intellectualize," I pick anything that looks like blueberries or huckleberries.

Blueberries and huckleberries are small (1/2 inch), blue to deep purple to black berries. The **highbush blueberry**, *V. corymbosum*, is probably the commonest blueberry of our New England states, the berry dark blue with a bloom. Despite proper nomenclature, the **huckleberry**, **dangleberry**, or **sunberry** may also be any one of several *Vaccinia*, and in fact the word *huckleberry* is a corruption of **whortleberry**, in turn a name for the *blueberry*.

Huckleberries, (or are they blueberries?) in Virginia, at least, grow sparsely in sandy soil, on bushes 12 to 18 inches high. In general, the places to look for blueberries and huckleberries (also charmingly called **wonderberries**) are in open fields, in any area that has been recently burnt over, along country roads, and on sandy paths at the edge of sparse woods. They grow only in wide, open scrubby areas, and they like light soil and sunshine all day long. Euell Gibbons suggests that, when picking them, you lay a blanket or sheet beneath a plant and shake the

Blueberry Coffee Cake

2 cups biscuit mix
1/2 cup sugar
1/3 cup safflower oil
2 eggs
1 cup milk

Preheat oven to 350 degrees. Mix dry ingredients thoroughly. In a separate bowl, beat liquids together. Mix the two lightly, and pour into a square nine-inch pan. Add topping, which consists of 1/2 cup sugar, 1/2 teaspoon cinnamon, 1/2 teaspoon allspice, and four tablespoons melted butter, tossed together with two cups blueberries. Bake until done, for about 45 minutes.

Blueberry Buckle

Cream four tablespoons of butter with 3/4 cup sugar. Beat in one egg. Add two cups of self-rising flour (or two cups of flour with two teaspoons of baking powder and 1/2 teaspoon of salt) alternately with 1/2 cup milk. Fold in two cups of washed blueberries. Pour into a greased, nine by nine-inch pan. Sprinkle the top with a crumb topping made by blending four tablespoons of butter, 1/2 cup of sugar, 1/3 cup of flour, and 1/2 teaspoon of cinnamon. Bake for 35 minutes at 375 degrees.

Wild Blueberry or Huckleberry Pie

6 cups wild blueberries
4 tablespoons cornstarch
1-1/4 cups sugar
2 tablespoons fresh lemon juice
1/2 grated nutmeg

Make two recipes of pate brisee or rich piecrust. Mix the above ingredients together, pour into one shell, dot the top with two or three tablespoons of butter, top with the second shell, and bake for 20 minutes in a fairly hot oven (about 400°F) until golden.

Bannock

2 cups biscuit mix
1 cup water
2 cups fresh wild blueberries or any
    berries of an edible sort

This is an old Girl Scout recipe. Make a dough, kneading lightly. Grease a hot skillet, put half the dough on the bottom, then the berries, then the rest of the dough. Cook over an open fire, turning when the bottom layer seems done. Top with butter or bacon grease (and sugar if desired) for a great breakfast.

berries into your blanket. It sounds efficient, so I pass the idea along.

EDIBILITY AND RECIPES

*F*ortunately, all varieties of both plants are edible and good, so no harm can come of picking whatever looks like a blueberry or huckleberry.

ETYMOLOGY

*B*lueberries and **bog cranberries** belong to the genus that the 18th-century plant classifier Linnaeus called *Vaccinium*, after the cows that probably grazed near—and sometimes on top of—the fruits. Foxes must have been observed eating cranberries and/or blueberries at some time, as a nickname for both is **foxberry**. The valuable and delicious berries go by a confusing array of folk names. For instance, the **farkleberry** or **sparkleberry** is a southern variety, *Vaccinium arboreum*, an evergreen small tree with a black, hard, dry berry resembling a huckleberry. *V. melanocarpum*, *V. stamineum*, and *V. caesium* are all called **deerberry** and **squaw-huckleberry**; all are species of *Vaccinium* "with no seeming constancy in shape and size of leaves or color and flavor of fruit," according to *Gray's Manual of Botany*. The English **bilberry**, *Vaccinium myrtillis*, now happily naturalized, is also known as **buckberry, American huckleberry,** and **whortleberry.** *V. myrtillis* also goes by **bear huckleberry, squawberry, hurtleberry, squaw huckleberry,** and **deer berry.**

HISTORY AND FOLKLORE

*J*ohn Josselyn, that curious curate who visited New England twice, in 1638 and again in 1663, related how the Indians of New England dried "bilberries" and sold them to the Englishmen who put them into "Puddens." Bilberries were good for "Feavers and Agues, either in Syrup or Conserve."

REPUTED MEDICAL VIRTUES

*H*istorically, bilberry leaf tea, like *Rubus* leaf tea, has been used to treat diarrhea and urinary infections, because the folk intuited a truth: that what is astringent tends also to be antibiotic.

The bilberry of England, *Vaccinium myrtillis* (but not the American species, *V. myrtilloides*) contains vegetable compounds called *anthocyanocides,* which have proven remarkably effective in increasing night-vision. It came first to the attention of British and American pilots in England during World War II. The aviators found that, after eating the tasty bilberry jam common to English households, they could see much better on their night flights. The effect would wear off, however, after five or six hours. The fresh fruit contains only a little of the anthocyanicides, but the dried or concentrated (as in jelly) berries have much more.

Quick Blueberry and Currant Tarts

1 quart blueberries
12 baked tart shells
1 pint currant jelly
sweetened whipped cream with a drop of vanilla

Melt the jelly over hot water or in a microwave. Distribute the berries in the tart shells, pour the jelly over as a glaze, and chill until serving time. Decorate with whipped cream at the last minute. This works beautifully with wineberries, raspberries, or black raspberries. I've done it more times than I can recall, and it's among the freshest desserts I make.

Today, herbal remedies including at least 25 percent anthocyanocides are catching on fast. It has been discovered that a pigment called anthocyanin also has impressive effects on capillary microcirculation. The implications of this are far-reaching; it may be that bilberries can safely and effectively reduce the edema of pregnancy, hemorrhoids, varicose veins, and so on by preventing capillary leakage. Because anthocyanin is also anti-inflammatory, it is being considered for use topically as a reducer of acne and other skin inflammations. It has proven effective in preventing cataracts in the elderly, and it is potentially useful in other eye ailments.

Betty Letcher's Huckleberry Cake with Hard Sauce

1/3 cup butter
1 cup sugar
2 eggs, beaten
1/2 cup milk
1-3/4 cups flour
1/2 teaspoon salt
1 teaspoon vanilla
2 teaspoons baking powder
1 to 2 cups huckleberries or
    blueberries

This is the favorite summer recipe of my childhood. Cream the butter, adding half the sugar gradually. Beat the remaining sugar with the eggs until thick and yellow. Combine mixtures. Add vanilla to milk. Mix and sift flour, baking powder, and salt, and add alternately with milk mixture to the first mixture. Beat well. Flour berries by tossing them in 1/8 cup of flour until all are coated. Pour into a sieve and shake gently to remove extra flour. Fold into batter. Bake in a loaf tin at 375 degrees for about 45 minutes or until done. Serve hot, in slices, with cold hard sauce. (You could substitute a good yellow cake mix, I'm sure, with results almost as good.)

Hard Sauce

5/8 stick of softened butter
3 tablespoons cream
1 tablespoon brandy or bourbon
1 tablespoon rum, dark
1 tablespoon sherry
1/2 teaspoon vanilla
1 pound powdered sugar, more or less
1 nutmeg

No substitutes allowed here. Cream the butter, slowly adding the liquid ingredients, then bit by bit the powdered sugar, until the mixture is rich, smooth, and homogenized. You can do this in a blender. Mound it in a dish, grate half a fresh nutmeg over the top, and chill until serving time.

Cranberries

DESCRIPTION AND LOCALITY

*B*og cranberries (*Vaccinium Vitis-Idaea* or *V. macrocarpon*) are the wild cranberries we're most familiar with. They are a low, creeping, evergreen shrub, generally no more than three inches high. Both the leaves and the berries tend to be small—the fruit smaller than cranberries found in markets. Cranberry plants have skinny branches, the fruit-bearing branches erect but still low. The spring-to-summer flowers are white to pinkish and bell-shaped, and the plant usually occurs where water is within root-reach. Cranberries prefer sandy soil and protected locations, and grow most abundantly from Cape Cod south to New Jersey and across to Wisconsin. The most likely place to find them is along the borders of ponds, down low where they can find protection from freezing, with the berries growing by the water's edge. Cranberries do fine with their feet in bracky water, ripening in fall and on into winter. The Indians showed Europeans the way to harvest the small berries—directly into their canoes, using rough wooden rakes like big combs.

A similar species, *Vaccinium erythrocarpum*, is also called **bearberry**, **American cranberry**, **lingberry**, **dingleberry**, and **Southern mountain cranberry**. It grows in the high mountains, in boggy areas of Virginia, Georgia, and North Carolina. Its leaves are deciduous, oblong, bristly serrate, and the shrub is two to four feet high. The flowers are pale pink, the red berries of inferior quality. They're easily gathered by holding a wide basket under a branch and stripping off the berries.

Bog Cranberries

EDIBILITY AND RECIPES

The wild fruits taste identical to the cultivated variety, which is to say they are sour. The Indians enjoyed drinking cranberry juice and cranberry tea, and eating cranberry bread. They showed colonists how to make cranberry sauce or jam (sweetened with maple sugar) to accompany venison. We still make it about the same way today: Boil for ten minutes (until the popping stops); use four cups of cranberries, two of sugar, one of water.

ETYMOLOGY, HISTORY, AND FOLKLORE

The cranberry was called "crane-berry," if the lore is to be believed, because the flower drooped and bobbed like a crane's head. Or because the fruit stem is crooked in the manner of a crane's bill. Or, if you prefer, because cranes were fond of the fruit. (Probably other birds, with little to eat in the wintry New England landscape, were, too, but cranes could wade into the water and eat the berries off the drooping branches.)

Bog cranberries go also by the names **cowberry, mooseberry, squashberry, lowbush cranberry, lingonberry** (a Scandinavian name), and **bearberry.** One might imagine the mountain bears tangling with the hogs of the Appalachian settlers over the ripe fruit, for another folk name is **hog cranberries.**

The colonists must have been gratified by how much *larger* the native American berries were than European cranberries. Cranberries may have been one of the first things settlers could brag about to the folks back home in Europe.

Cranberry Tart

Take half a pint of cranberries, pick them from the stems and throw them into a saucepan with half a pound of white sugar and a spoonful of water. Let them come to a boil, then return them to stand on the hob while you peel and cut up four large apples. Put a rim of light paste around your dish. Strew in the apples, then pour the cranberries over them. Cover with a lid of crust and bake for an hour. For a pudding, proceed in the same manner with the fruit, and boil it in a basin or cloth. (Taken from *The Godey's Lady's Book Receipts and Household Hints* by S. Annie Frost, 1870.)

Cranberry Drink

4 cups cranberries
2 cups sugar
6 cups water

Cook berries, sugar, and one cup of water until all berries are popped. Add the rest of the water and let it sit for four hours. Strain through cheesecloth into jars.

Cranberry Stuffing for Turkey

2 cups cranberries
2 cups chopped onion
2 cups chopped celery
2 tart apples, chopped
1 stick butter
1 small package of bread or cornbread stuffing mix
1 cup water

This is easy. It's my own recipe, and it should stuff a turkey of about 15 pounds. Just melt the butter in the water and stir everything else in. It makes a tart and wonderful stuffing.

Bradford and the Mayflower Pilgrims anchored in Cape Cod Bay on November 11, 1620, but it was a month before they were able to move passengers onto land. Before that, skiffs of men went ashore to find water and food, and to assess the possibility of shelter and the neighborliness of the natives. Arriving on land as they did in December, the English travelers were, after three months at sea with little fresh food available during their journey, a scurvy lot.

By the sixth of April, 1621, when the Mayflower returned to England, half the company had died of dysentery, starvation, pneumonia, and nutritional deficiencies. Accounts from that time indicate that the toll would have been far worse had the Indians not shown the scurvy Englishmen a tea made from the light green, new-growth tips of evergreen limbs and a certain healing red berry.

After their salvation, the New England settlers systematically gathered the pretty red fruit, providing ships' crews with dried cranberries for voyages across the Atlantic over the next two centuries. Wild cranberries soon became a major money crop in New England and, of course, continued to provide one way of getting through the winters in good health.

The Algonquins made what must be a contender for the first fast food: pemmican. These natives were seasonally migratory, and pemmican was their traveling fare, consisting of ground cornmeal, dried salted venison, bear fat or some other suet, and dried cranberries or blueberries (or even serviceberries), all made into little patties and dried in the sun or baked on fire-heated rocks. Note that these little patties were balanced powerhouses of nutrition:

They contained meat, fat, grain, and fruit, a small amount of sugar from the berries, plus the concentrated Vitamin C of the dried sour berries. These journey-cakes would withstand rough treatment, travel efficiently, and resist spoilage.

Throughout the U.S., cranberries symbolize Thanksgiving as surely as turkey does; they have graced our harvest tables ever since the Indians introduced the fruit to Englishmen.

REPUTED MEDICAL VIRTUES

*T*he Indians' use of cranberries included poulticing wounds with the roasted berries. Blueberries and cranberries were both believed effective against poison-arrow wounds—and perhaps they were, as the berries are astringent, especially when raw. Cranberries have been praised for easing cramps and childbirth, even convulsions, "hysteria," and "fits." Other early writers noted that cranberry tea helps one overcome nausea.

RECENT UPDATE

*I*t has long been a tenet of folklore that cranberry juice is useful in curing chronic bladder infections, including persistent cystitis. Recent research confirms the folk knowledge; most chronic UTI's (urinary tract infections) are caused by *Escherichia coli* bacteria that live in the human intestines; certain members of the *Vaccinium* species contain *arbutin,* a substance that prevents the *E. coli* bacteria from adhering to the walls of the bladder or urinary tract, thus ridding the body of the germ. It takes a fairly long time and a lot of juice, but it's likely that even stubborn cases will yield in the end.

Raw Cranberry Sauce with Orange

4 cups cranberries
1 large orange
2 cups sugar

Clean the cranberries. Cut the orange into chunks, removing all the seeds carefully, but leaving the skin on. Grind the orange or quickly chop it in a food processor. Quickly chop the cranberries also. Stir both together and add sugar, stirring thoroughly. Refrigerate for three days before serving.

Highbush Cranberries

INTRODUCTION

*H*ighbush cranberry is a name given not only to the *Vaccinium oxycoccus*, but also to *Viburnum opulus*.

DESCRIPTION AND LOCALITY

*T*he *Viburnums* comprise a group of shrubs or small trees mostly with straight stems (thus the common folk name of arrow-wood), which are mainly found in the northern hemisphere. They have white or pink, perfect, five-lobed flowers mostly in flat umbels, which are also called *cymes*, (flowerlets in a flattish group with the central flowerlets blooming earliest). The **blackhaw**, or **southern blackhaw**, is one of two *Viburnums*, *V. prunifolium* or *V. rufidulum*, that is also sometimes called **rusty nannyberry** or **sheepberry**. *V. prunifolium* is a spreading large shrub or tree six to twenty feet tall, with smooth, pale gray branches, and toothed, thick, elliptic to oval leaves that are rusty-looking on the undersides. It grows in wet places throughout Canada and the eastern part of the United States, though not in the deep South. Its flowers are white in April or May; the berrylike drupes, called haws because they resemble the fruit of hawthorns, ripen in early autumn from green to red to bloomy black. They are usually 3/4 inch, more oval than round, and dull blue-black with a sweet pulp when ripe.

Highbush cranberry is also known as **crampbark** (from the plant's cathartic properties), **upland** or **mountain cranberry**, **squawbush**, **bearberry**, and **craberry tree** (and in England, **fenberry**). All the cranberries are said to yield a lovely, pink-red dye.

EDIBILITY

*S*ome of the *Viburnums* have edible, berry-like fruits that are actually drupes, roughly 1/2 inch to 1 inch, roundish in appearance like the *Vacciniums*. *Viburnum lentago*, with a berry about half an inch long, oval to almost round and bloomy blue-black when ripe, is variously called **sheepberry, highbush cranberry, nannyberry, sloe, black haw**, and **wild raisin**. The fruits ripen in autumn.

V. edule, a West Coast and northern *Viburnum*, is a scraggly shrub one to three feet high, with orange or red half-inch, roundish drupes; its leaves are sycamore-like. It makes a good jam. It is also called **mooseberry, squashberry, and bearberry**. *V. rufidium* tastes much the same as the others, and all are known by the same group of folk names. They all have a clean, musty odor, and are benign.

REPUTED MEDICAL VIRTUES

*C*ountry folk through the ages have made the inner bark of several *Viburnum* species into a tea for cramps and muscle soreness. Asthma among early settlers was alleviated by a dried bark tea of the *Viburnum opulus*. *V. trilobum* is currently used as an antispasmodic. Its leaves are large, wavy-edged, and three-lobed, with the lobes pointed. The roundish, translucent red drupes, under 1/2 inch, ripen in the fall.

The bark of *V. prunifolium* is made into a tea and used as a tonic. The Ojibwas used nannyberry for a diuretic tea, scraping the inner bark into boiling water. The fruits were a popular food with many Indian tribes. John Brickell, in his *Natural History of North Carolina* (1737), described how the bark of the black haw "being dryed and made into a fine Powder, and apply'd to inveterate old Sores especially in the Legs very speedily cleanses and dries them up, and is one of the best Remedies on these occasions, I have ever met with."

RECENT UPDATE

*H*ighbush cranberries have traditionally been consumed to overcome the difficult breathing of asthma. We have noted earlier that anthocyanins may help circulation by relaxing capillaries, so perhaps highbush cranberries really do dilate bronchial tubes, along with blood vessels. Tests to determine their efficacy are currently underway.

Gooseberries

The barbery, respis, and goosebery to,
Looke now to be planted as other things do;
The goosebery, respis, and roses al three,
With strawberies under them, trimly agree.

—*Thomas Tusser, in Five Hundred*
Points of Good Husbandry, 1597

DESCRIPTION AND LOCALITY

Gooseberries usually grow on prickly deciduous bushes with solitary or few flowers, and the pedicels or stemlets are not joined. Gooseberry bushes grow in graceful, waist-high arches, and the best-tasting ones have thorns. The ripe berries, as pretty to look at as rare marble, may be reddish, pinkish, yellowish, purplish, brownish, greenish, scarlet, midnight blue, stripish, or whitish with striations (like small basketballs). Some of the berries are hairy or prickly.

If you are inclined to search gooseberries out in your part of the country, study the illustrations to learn their distinctive leaves and markings. The berries crop up wherever the climate is coolish, but they are more common in New England than elsewhere. However, there is a species, *Ribes curvatum*, with white flowers and smooth, greenish fruit that grows in the deep South. Among the commonest in this country are the native *Ribes cynosbati*, the **wild gooseberry** or **dogberry**. The flowers are light green, and I would describe the color of the prickly fruit as reddish to brownish-purple. *R. hirtellum*, the **smooth gooseberry**, common in all the northern states and Canada, has purplish flowers and deep purple berries. *R. lacustre*, the **swamp gooseberry**, has purplish-green flowers and bristly purplish fruit, and a good flavor when cooked. It exudes a bad smell when raw, though, which leads to the unfortunate nickname of **skunkberry**. *R. oxyacanthoides*, the **northern gooseberry**, has greenish flowers and glabrous (smooth, hairless) reddish fruit; it grows throughout

Canada and Newfoundland all the way to the Arctic circle. *R. divaricatum* of the Pacific northwest, the **western gooseberry**, has smooth, dark purple berries. A number of species of *Ribes* are host to a blister rust that is deadly to the white pine, so they are not welcome in states where white pine is a cash crop or a valued resource.

EDIBILITY AND RECIPES

*T*he fruits are small, many-seeded, pulpy, and often sour. Nonetheless they have enjoyed modest popularity as candidates for pies and preserves. Gooseberries are as useful as cranberries, and are high in ascorbic acid. They immigrated to America from England, but were plagued by mildew from the beginning and never became as popular as other common berries. This may be because their taste and consistency are so variable, and because of limited availability. By the time a mildew-resistant strain of gooseberries was developed in the mid-19th century, no one took much interest.

I once tasted a turkey stuffing with gooseberries in it; the tart taste was pleasant amid the herby, oniony, and bready flavors. I have made gooseberry pies, but only twice. The first time, I found the taste so uninteresting that the second time I added allspice and cinnamon. Still, I concluded (apparently along with most Americans) that it was not a fruit I'd pursue.

ETYMOLOGY, HISTORY, AND FOLKLORE

*R*ibes, a cognate of *Ribas*, is the name given to the genus in the 11th and 12th centuries by Arabian medical botanists. In the British Isles, gooseberries go by many folk names: **peaberries**, **carberries**, and **thapes** among them. No one is sure where the silly name of gooseberries came from, though perhaps it derived from

Gooseberry Chutney

1-1/2 quarts gooseberries
1 cup raisins
1 big onion, chopped
2 cups brown sugar
3 tablespoons dry mustard
3 tablespoons powdered ginger
3 tablespoons salt
3 cloves of garlic, chopped fine
1 tablespoon mace, ground
1 quart cider vinegar

This is delicious. The garlic is my addition to an old recipe. Simmer together over low heat for an hour and a half, then bottle and seal. Serve with any meat, but especially roast lamb.

Gooseberry Fool

1 quart gooseberries
1 cup cranberries or other red berries
1 cup sugar
juice of one large lemon
1/2 cup sour cream
3 tablespoons sugar
1 teaspoon vanilla
1 cup whipping cream

Simmer until the berries pop, with 1/2 cup sugar, lemon juice, and two tablespoons of water. Taste, and add more sugar if needed. Mash the mixture some and chill. When it is cold, whip the cream with sugar and vanilla until soft peaks form, fold in the sour cream, then the berries. The red berries tint the beige gooseberries a delicate pink color.

the berry's association, more or less from the beginning, with roast goose. One theory suggests that the plant's resemblance to the gorse, or whinbush, led it to be called "gorseberry," then eventually, through misunderstanding, gooseberry.

In England, gooseberries are almost a national symbol. In some counties in the 19th century, annual summer gooseberry-prize meetings were held, with the heaviest berries claiming the top awards. In fact, an annual book of the winners was printed, called *The Manchester Gooseberry Book*. Gooseberry wine remains popular in England, and Gooseberry Fool is a favorite dessert.

REPUTED MEDICAL VIRTUES

*G*ooseberries are regarded as mildly medicinal, generally useful for stomach complaints. A tea made from both fruit and leaves was used by some Rocky Mountain Indians for prolapse of the uterus, a condition caused by too many pregnancies. In colonial days, gooseberry juice was used internally as a foolproof cure for colds and sore throats—probably it was the high vitamin C content the pioneers instinctively sought—and externally by certain western Indian tribes to soothe skin irritations like poison ivy and erysipelas, a streptococcus infection characterized by inflamed skin and mucous membranes accompanied by systemic fever.

Currants

DESCRIPTION AND LOCALITY

*W*ild currants grow all over the northern hemisphere. **Currants** are often without spines at the nodes, and the flowers (and subsequently the fruit) are mostly racemes (clusters of single flowers on small stems growing at intervals along a single larger stem). The flower color, like the berry color, varies considerably, ranging from greenish-white to pink, red, purple, yellow, or brown. Currants have lobed leaves on long stalks. Ripe red currants, hanging gracefully with the sun shining through them, are gorgeous. **Black currants** are a distinctive variety, whereas white and "champagne" currants are variations of *R. rubrum*, the **red currant**.

EDIBILITY AND RECIPES

*C*urrants have a tart taste not much suited to eating raw. Because the flavor of the most popular, the **wild black currant**, is "skunky" when eaten uncooked, it has been dubbed **skunkberry** by some. But currants make absolutely wonderful jelly with a rich, deep, winey taste— among the most popular varieties of jelly in Europe. Currants are the base of Creme de Cassis, a fruit liqueur. **Jostaberries** are a popular new European cross between black currants and gooseberries; the resulting delicious jelly is called, appropriately enough, Gelatina di Ribes.

ETYMOLOGY, HISTORY, AND FOLKLORE

*T*hey are called currants by way of a corruption of *Corinth*, a city in Greece, in whose vicinity wild red grapes once grew. Those grapes, not currants, were dried into—well, currants of a sort—that became famous throughout Europe in the Middle Ages.

Red currants (*Ribes vulgare*) may have red or white fruit and can be used as the gooseberries are, or in jelly, sauces, or wines. Laws prohibit growing *R. nigrum*, **European black currants**, in some states in this country, as they are hosts for a certain stage of white pine blister, which kills those handsome trees.

Currant or Raspberry Catsup

(The condiment's name is from the Chinese ke-tsiap, a fish sauce brought to Europe by English and Dutch traders from the Orient in the 18th century. In the 19th century, catsup was made from blueberries, cranberries, and even from oysters, before someone decided that tomatoes were the only appropriate fruit for such a condiment. Taken from *Household Discoveries*, 1908.)

5 lb. gooseberries or currants
4 lb. sugar
2 cups cider vinegar
2 tablespoons ground cinnamon
1 tablespoon ground cloves
1 tablespoon ground allspice
1 teaspoon ground mace
3 tablespoons salt

Clean fruit. Put into enamel kettle with other ingredients and simmer gently for two hours. Fill bottles and seal.

Currant Salad

Take a good quantity of blanched almonds, and with your shredding knife cut them grossly; then take as many raisins of the sun, clean washed and the stones picked out, as many figs shred like the almonds, as many capers, twice so many olives, and as many currants as of all the rest, clean washed, a good handful of the small tender leaves of red sage and spinach. Mix all these well together with good store of sugar, and lay them in the bottom of a great dish, then put unto them vinegar and oil, and scrape more sugar over all. Then take oranges and lemons, and paring away the outward shells, cut them into thin slices. Then with those slices cover the salad all over; which done, take the fine leaf of the red cabbage and with them cover the oranges and lemons all over. Then over those red leaves lay another course of old olives, and the slices of well-pickled cucumbers, together with the very inward heart of cabbage-lettuce cut into slices; then adorn the sides of the dish and the top of the salad with more slices of lemons and oranges, and so serve it up. (Taken from Gervaise Markham's *The English House-wife*, 1615.)

Currant and Raspberry Paste Drops

One pound of fruit, either kind or a mixture of both. Boil and press the pulp through a fine sieve. Add an equal amount of sugar. Boil down until a small bit dropped in cold water will harden. Drop quickly onto a lightly greased pan by teaspoonfuls. Let stand for half a day to dry, then wrap each in a square of waxed paper and store in a dry place.

REPUTED MEDICAL VIRTUES

*C*urrant juice drunk regularly has long been said to alleviate the joint pain of arthritis. Certainly the berries are antiscorbutic. Eskimos eat currants as insurance against the heart disease that can result from their fatty diets. A leaf infusion of currants is said to be cleansing and diuretic. Certain Indians used the plant roots boiled in water to allay kidney problems.

RECENT UPDATE

*M*odern investigations of the medicinal qualities of currants and gooseberries indicate that both manifest strong antiseptic properties, have proven effective against *Candida* infections, and have even shown promise in some cancer treatments. *Ribes* contains Gamma Linolenic Acid (GLA), which occurs rarely in nature; the best sources thus far identified are evening primrose and human milk. What makes a baby strong may not do a thing for an adult, but proponents of GLA believe it can strengthen the immune system and the heart, improve circulation, offer relief for migraines, and mitigate menstrual and pre-menstrual problems.

Raspberry and Currant Punch

1 pint raspberries
1 quart currants
    Bruise the fruit in a kettle and pour 1/2 gallon of water over it. Heat the mixture slowly and bring exactly to a boil, then remove it. Pour contents into a jelly bag and let it drain into a large bowl. Sweeten it with sugar to taste, and add ice cubes for a pleasant drink.

Mulberries

DESCRIPTION AND LOCALITY

Mulberry trees are variable in size and foliage, and are long-lived. They are 30 feet tall on the average, though red mulberries grow as high as 75 feet in the South. The large (two to eight inches long) leaves of mulberry trees are thick, dull, and dark green. They are toothed, often lobed, and the trees have a milky sap. The multiple compound fruit, each bead fertilized independently, is ovate, compressed, and covered by the succulent calyx. The whole spike (stalkless flowerlets on an elongated common stem axis) becomes the edible fruit. Mulberry trees have spread from early cultivation in Virginia and other eastern states.

The hardier black or red varieties of mulberry grow along city streets and back alleys, in meadows, along roadsides, in back yards from New York to the Carolinas, and west to Missouri.

EDIBILITY

The aggregate fruits, which are not true berries, ripen in early to late summer. Because red mulberries are acidic, they are tastier than the black or white varieties. Southerners have long made a popular cheap wine from red mulberries, but *M. rubra* is now primarily considered important as a food for wildlife. It has been noted that the taste of mulberries varies enormously from species to species, and even from plant to plant. Some are sweet, certainly, but flat—sweet without any tang.

Children eat them at times, but most often the ground beneath a tree of ripe mulberries will be red with trampled berries and their juice. Nobody picks up the soft, juicy berries, or climbs to pick them. With the addition of lemon juice, however, they might be turned into marvelous jellies, jams, and pies. Sometimes it seems that we're not hungry enough these days.

ETYMOLOGY, HISTORY, AND FOLKLORE

A charming Greek legend says that the berries of the white mulberry turned red when its roots were bathed in the blood of Pyramis and Thisbe, lovers who committed suicide because each mistakenly thought the other was dead. (Shakespeare, as you may recall, parodied and revived the legend in *A Midsummer Night's Dream*.)

The Romans, who wanted everything the Greeks had, also wanted silk. In 220 A.D. Heliogabalus, Emperor of Rome, obtained and proudly wore a silk robe. So did Aurelian, a generation later. The Romans, again copying the Greeks, imported from Persia and cultivated the black mulberry (*M. nigra*) as a host for their own silkworm cultures.

In America in 1623, the Virginia House of Burgesses required that each free man must cultivate one-quarter acre of "vines, herbs, and roots." A year later, the law became more specific: Four mulberry trees and 20 [grape] vines had to be planted by every male over the age of 20. The Virginia Company, its economic venture failing after only one generation, was determined to recoup its losses through the wine and silk industries. Neither thrived, though the law stayed on the books for many years. No matter how much silk was made, better could be purchased cheaper from the Orient, where labor was almost free.

REPUTED MEDICAL VIRTUES

A exander of Athens wrote in 340 B.C. of a blight of mulberry trees in his time so great that for 20 years there were no mulberries. Gout raged, he reported, so that men, women, children, and eunuchs all got the disease. Obviously he considered

mulberries an effective cure for gout.

On this continent, American Indians used the native red mulberry as a mild laxative, and a drink from the fruit has been used to allay fever. Syrup of red or white mulberry is employed in Appalachian folk medicine to soothe coughs. Red mulberry root bark has been used as a vermifuge—that is, to rid the body of tapeworms and other intestinal parasites. It is shaved, pounded, then boiled as a tea and drunk. The tree's sap, applied directly to the affected area, was used by Indians and white folks alike in treating fungal skin ailments such as jock itch and ringworm.

Like blackberries, mulberries are still used in Appalachia to treat gout. When my grandfather suffered from gout, an herbal doctor advised mulberry juice and told him, "Hogmeat done it." Unhappily, my grandfather abandoned ham and bacon. Today we know that gout is hereditary, but it can be somewhat mitigated by whatever methods can flush uric acid out of the body.

INTRODUCTION

*S*erviceberries (pronounced "sarvis" in Appalachia, as it would have been pronounced in Shakespeare's England) are so-named because they *look* like berries (they aren't) and because the distinctive white flowers in the still-barren woods signal the end of winter—the time when snowbound circuit-riders would begin to make their rounds in the unpaved southern mountains, attending to the considerable chore of burials from the past winter. An old mountain man once described to me his boyhood memory of great piles of coffins in the awful spring of 1919, the boxes awaiting the "services" of the preacher in the mountains near the Virginia-Kentucky border.

DESCRIPTION AND LOCALITY

*S*erviceberries, the fruits of several species of trees in the *Amelanchier* genus, can be marked for June or July picking by noting the pure white, five-petalled blossoms in narrow clusters early in the spring—even before the dogwoods and redbuds bloom. This beautiful, small, deciduous tree can be found in woods everywhere in the northern hemisphere. The medium-sized leaves of *Amelanchier* are coarsely or finely-toothed, and the bark is smooth and gray. Each dry, insipid, sweetish berry contains ten seeds. Though often called **Juneberries**, one species, *A. bartramiana*, which grows as far south as Virginia, blooms from May to

August. The fruit ripens in September.

Serviceberries are tiny pomes. Among the many species and sub-species, leaf shape differs significantly. *Gray's Botany* accuses them of "commingling and producing many perplexing hybrids and mongrel offspring," and lists 19 species, admitting to innumerable varieties or sub-species. Service trees have a graceful shape, lovely white spring flowers, bronze leaves in the autumn, and dark blue, edible berries the size of blueberries in between. You can't ask much more of a plant than that. It's popular today as an ornamental tree. Plant it if you like to feed wild birds.

EDIBILITY AND RECIPES

To the Indians, serviceberries were a valuable spring food. They were often dried like raisins, and the berries have been described as juicy and sweet. All are tiny, about one centimeter in diameter. They are nearly always black-purple (although one has white fruit).

Serviceberries are all edible, but the quality varies a great deal. The **saskatoon** (*A. alnifolia*), named by the Indians and used long before white men came, is apparently one of the better ones, as it is grown in Saskatchewan for commercial distribution. The berries, if you like the taste, can be used in pies, and can be frozen or made into a sauce. You could even try pemmican, that famous Indian snack, and take it along to munch on as you seek fresh berries in the woods. It might make the other berries you find taste even better.

Serviceberry Preserves

2 cups water
4 cups sugar
6 cups serviceberries
3 tablespoons lemon juice or vinegar
1 package pectin

Boil the water and berries for ten minutes. Drip and press through cheesecloth to make four cups of juice. Mix juice and pectin, bring to boil, and dump the sugar in all at once. Boil furiously for one minute, then remove from the heat, skim, and jar.

Serviceberries are known in the Chesapeake Bay area as **shadbush**; in New England as **shadberries** or **shadblow**; in Canada as **pimbinas**; and elsewhere as **sugarpears** or **Juneberries**. To the native Americans, the bright white (in some species, pale pink) blossoms in the gray, wintry woods were easy to spot. They meant that winter was broken and that fresh fruit would soon be available. The *Amelanchiers* are what nurserymen today call *indicator plants*: When serviceberries bloom, the shad also run up tidal rivers from the sea to spawn. Knowing this gave the Indians a decided advantage over the shad—hence the many nicknames associated with that delectible fish. Most trees bloom when they get ready and don't necessarily coincide with anything else.

RECENT UPDATE

The *Amelanchier's* blooming date in any given locality directly correlates with the emergence of several scale-insects, not to mention the dogwood borer and the bronze birch borer. Knowing this enables nursery workers to treat plants specifically to control the later emergence of various insects and spiders.

Pemmican

Mix a pound of dried beef or buffalo, a half pound of dried serviceberries, and a pound of beef, bear, or buffalo suet with as much ground Indian meal (cornmeal) as the mixture will hold. Form it into patties. Dry them in the sun, on a hot rock, or in the coals of a fire.

Serviceberry Pie

1 unbaked crust
3 cups serviceberries
2 tablespoons flour
1 cup sugar
1 tablespoon butter
1 tablespoon lemon juice

Mix ingredients, fill shell, and bake.

Elderberries

DESCRIPTION AND LOCALITY

*E*lderberries are a folk favorite, widely and safely used, especially in Appalachia. Yet at certain stages, elderberries are poisonous.

Elderberry species grow throughout the United States, including Hawaii and Alaska, and are members of the honeysuckle family. The leaves are compound and opposite, with five to eleven pointed and serrate leaflets. Elderberry grows in dense thickets, its off-white flowers borne in a flattish bunch, rather like Queen Anne's Lace. The shrub yields small clusters of blue-black fruit in late summer. The elderberry is actually a drupe, black when ripe and the size of a small pea.

Elderberry is any one of a dozen perennial, deciduous shrubs, six to ten feet high, of the genus *Sambucus*. Their many-branching stems are so pulpy that they have been hollowed out and used as blowguns and syringes in times past. The red-fruited species (*S. racemosus*) with the berries in domes is poisonous. In fact, consider all parts of all the species poisonous, except the flowers and ripe fruits.

EDIBILITY AND RECIPES

*E*lderberries are an excellent fruit for jelly or wine, with an intense, spicy-grapey flavor *when cooked*. They are ripe in August or September. Elderberries and their flowers are still made into wine today by country folks here and in Europe, and an Italian liqueur called Sambuca is flavored with the berries. Elderberry pie is common in Appalachia, and no one suffers ill effects from eating it. In spring, the white flower heads are dipped into batter and fried into fritters—hence the folk name "**fritter tree**." Ripe elderberries are supposed to be high in vitamin A, so the plant can hardly be called poisonous.

But the plant *is* poisonous. The leaves, stems, green berries, and roots contain cyanogenetic glycosides as well as an unidentified cathartic. Eaten unripe or uncooked, elderberries cause vomiting and diarrhea.

ETYMOLOGY, HISTORY, AND FOLKLORE

Sambucus was named for the *sambuke*, an ancient Greek musical instrument made from the hollowed stems of elderwood. Some say sambukes were the original pipes of Pan. The cross on which Christ was crucified was supposedly made of elderwood; thus the plant is reviled by some. Yet in the 17th century, the English writer John Evelyn proclaimed elderberries a "remedy against all infirmities whatever."

Folklore credits the elderberry plant with magical properties, among them the ability to confer invisibility and second sight, and the power to cure the bites of mad dogs. Elderberries were believed by the 16th-century English herbalists Culpeper and Gerard to restore youth; perhaps this is because gray hair could be returned to black with a dye made from the elder. The berries themselves are used to obtain crimson or violet dye. Elderberry leaves yield a clear, yellow-green dye. The berries are by most sources labeled a poison, though that is only part of the story.

Battered Elder Blooms

1 cup flour
1 teaspoon salt
1 tablespoon sugar
1 tablespoon oil
1 egg
1/2 to 3/4 cup of beer, or enough to make a fairly thin batter

Mix the batter as above, chill it well, and dip clean, dried elderberry blooms into it. Fry them in deep fat preheated to 360 degrees.

Elderblossom Wine

2 quarts of packed elderberry blossoms
10 pounds sugar
1 cake yeast
4 gallons water
1/2 cup lemon juice
3 pounds raisins

Take all stems and leaves out of the blossoms, wash them, and pack them into jars. Stir the sugar and water together until dissolved, then boil for five minutes without stirring. Skim, then add blossoms. Remove from fire at once.

When lukewarm, stir the yeast into 1/2 cup of the liquid, then into the lukewarm wine. Add the lemon juice and stir. Put in an earthen jar for six days, stirring thoroughly three times a day to bring the blossoms up through the wine. On the seventh day, strain through cheesecloth and add the raisins. Put in glass jars and cover tightly, but do not bottle until New Year's Day.

Elderberry Pie

1 quart or so of ripe elderberries
1 cup sugar or a little more
dash of salt
1 tablespoon cornstarch
2 tablespoons lemon juice
3 tablespoons butter
a double crust

Mix the first five ingredients together. Pour into one crust, dot the top with butter, and lay the other crust on top. Cut venting slits and bake until golden. This is a pie my grandparents' servant Lizzie used to make. It has a strong, winey taste, and ought to be served with ice cream on top.

Spiced Elderberry Syrup

4 quarts ripe elderberries
1 cup water
1 cup sugar
1 stick cinnamon
1 tablespoon whole cloves
3 small pieces of dried ginger root

This syrup is reportedly a mild laxative. Boil it for an hour gently. Then strain, mashing the pulp through the strainer, and refrigerate. Half a cup two or three times a day will cure constipation.

REPUTED MEDICAL VIRTUES

Dried-elderflower water is said to be good for sores, blisters, hemorrhoids, rheumatism, and arthritis, as well as for the maintenance of glowing complexions. Elderflower Under-Eye Gel and Elderflower Water are featured in cosmetic displays today. Skin "toners" and astringent lotions contain extracts of the plant.

A substance in the bark of the elderberry tree is an effective painkiller. Choctaw Indians pulverized leaves of elder with salt for a headache poultice; an inner bark tea was held in the mouth for relief from toothache. Bark tea was also used to clear the lungs of phlegm in cases of pneumonia and was used to cure colds.

A decoction of leaves and flowers was used to wash open wounds and keep maggots from laying eggs in them. The inner bark was used as a laxative, and the berries were said to be aperient, diuretic, diaphoric, and cathartic. They were believed valuable in treating "rheumatic gout," scrofula, and syphilis, and the juice was supposed to be "cooling."

Compresses of elderflower, mashed into boiling water and applied as hot as possible, were said to relieve the pain of hemorrhoids. In colonial days, elderberry flowers were made into a veterinary salve by pounding the flower heads with lard, heating and melting the mixture, and straining it into jars. It was said to rid animals of flies and fleas. The same ointment was used by discriminating ladies as face cream.

RECENT UPDATE

The elderberry plant contains substances that release cyanide and cathartic principles, which explains some of the plant's medical uses, but also indicates its dangers. The blossoms and ripe berries are safe; the rest of plant is not.

Otherwise Useful Berries

(including some pretty poisons)

*"There ain't no poison [that] ain't to
some a medicine . . ."*

—*Edgar Shew, Pocahontas, Virginia, 1982*

DESCRIPTION AND LOCALITY

*B*arberries are cool-weather plants native to every continent but Australia. On our continent, barberries grow from the Arctic Circle as far south as Delaware and Pennsylvania, and hardy garden varieties thrive outside this area. Related species grow in the South and the West. The plants are perennial shrubs that grow to about six feet in pastures, light woods, and fence rows. Many varieties are popular landscaping plants. Barberry leaves are small, oval, and grayish underneath, turning rusty red to winey in the fall. The flowers are yellow racemes, and the fruit is oblong, bright red to deep scarlet or purple. The plant has sharp spines at each node.

EDIBILITY AND RECIPES

*T*he **barberry** is listed as poisonous in many texts, but its folk names suggest otherwise: **mountain grape, oregon grape**, and **Rocky Mountain grape. Sowberry** suggests that wild pigs liked it. Many early cookbooks contain recipes using barberries. The roots, used for a bright yellow dye, contain dangerous alkaloids, but the berries are often used for jams, pickles, and jellies, so it is highly unlikely that they are toxic. Their taste is reminiscent of cranberries, and they are slightly but not unpleasantly bitter.

ETYMOLOGY, HISTORY, AND FOLKLORE

*O*ther names given to the barberry are **berberry**, and **jaundice berry**. The name derives from *Berberys*, the Arabic name for the fruit. They thrive throughout the eastern states, doing no harm as long as they don't occur near wheat fields. In states where wheat is a major crop, some species of the plant are outlawed because they harbor a wheat fungus.

REPUTED MEDICAL VIRTUES

*C*atawba Indians pounded the roots and stems of barberry and made a tea to soothe stomach ulcers. Penobscot Indians used barberry on sore throats and ulcerated gums. The fruit was supposed by one early observer to be "cooling in Fevers, grateful to Stomach, and causeth a good Appetite." Indeed, the fruit is rich in vitamin C. It was reportedly used by American Indians as an aperitif, and early settlers copied the natives by making of it a bitter tonic, which was taken to increase the appetite.

Large doses have been reported to have a cathartic effect. Modern pharmaceutical tests have not found much promise in barberry, despite many efforts and close investigation into its folk uses.

Barberry Jelly

Heat several cups of barberries slowly in a deep pan, mashing with a wooden masher and cooking slowly until the fruits are soft. Add a bit of water if the fruit is dry. Then strain the juice through double cheesecloth and measure. Put it back in the pot to boil. Meanwhile, measure 2/3 as much sugar, and add one tablespoon of cider vinegar for every quart of juice. When the juice boils, add sugar and vinegar at once, and boil rapidly to the jelly stage.

Pickled Barberries

Place the barberries, clean and stripped off the stems, into a jar of heavy brine with a clove of garlic and a pinch of pickling lime. In a week or so, they are ready to garnish cold meats or salads.

Juniper berries

DESCRIPTION AND LOCALITY

*J*unipers grow on dry hills, in old fields, and in rocky soil everywhere. The evergreen can be scrubby or narrowly upright. The bark is reddish, and the sharp, needle-like leaves are arranged in threes, averaging one-half inch long. The aromatic cone, erroneously called a berry, is oval, about a quarter-inch in diameter, and dark blue with a slight bloom. *Juniperus communis* is the commonest indigenous New England species of cedar tree, while *J. virginiana* occurs more often in the temperate zone. The western juniper, *J. occidentalis*, grows west of the Mississippi, as do many other species.

EDIBILITY AND RECIPES

*J*uniper occidentalis bears a large, sweet, nutritious blue cone that the western Indians gathered for winter. The flavor of the **cedarberry** or **juniper berry** is known well by those who drink gin, and the insect-repelling scent of the wood is familiar to everyone who has a cedar chest or closet. Today, juniper is used mostly as an aromatic rather than a food.

ETYMOLOGY, HISTORY, AND FOLKLORE

*T*he juniper's name suggests a mysterious ability to stop time, in that *Juniperus* means "youth-producing." Its folk name is **savin,** or **savine,** from an English juniper, *J. sabina*, which was much loved by English gardeners in previous centuries.

The juniper berry yields the fragrant oil favored by Englishmen and anti-prohibitionists. The perfumy wood is made into hope chests, with the promise that anything stored in them will be safe from depredation by insects. By analogy, the bride hoped that her future love and happiness would also remain fresh and unsullied. Juniper is mentioned in Columella's *Rural Economy* (76 A.D.), and it has been used medicinally since at least the Middle Ages.

The American Indians had many uses for juniper. They seasoned meat, salmon, and other foods with it, and they used it as a repellent for rodents and insects. The plant also yields a blue-green-gold dye.

REPUTED MEDICAL VIRTUES

A common nickname for the juniper is **medicine berry,** and a forbidding number of human ailments are said to respond to juniper's medicine: unwanted pregnancy, snakebite, colic, kidney stones, worms, burns, infections, flatulence, scurvy, dropsy, bronchitis, and insect infestations. It is said to "excite an organic action in the human system." Drinking juniper tea was supposed to "purify the blood." The Aztecs used a boiled tea of a Mexican juniper, *J. mexicana*, to treat skin infections, and the Plains Indians made a fire for those with chest infections by burning cedar twigs and enveloping the sufferer's head in a blanket so he could inhale the smoke. The same treatment was employed against evil dreams. In Appalachia, all parts of the plant are steamed as inhalers for folks with winter cold symptoms or bronchial problems.

Tricky Marys

Twice I have served spiced (but not spiked) tomato juice to large groups of people, with ice, celery, and olives, as Bloody Marys. What the guests did not know is that for three days prior to the parties, I had soaked juniper berries in the tomato juice. Everyone *thought* they were drinking gin, and both times some people experienced the lightheadedness that one would expect as a result of a cocktail. The first time was experimental: a brunch before a wedding. I thought it might not be especially helpful to the wedding party to serve alcohol at ten in the morning, so I invented this version of Virgin Marys. The second time was calculated. People again had a wonderful time, and there was no piper to pay.

Juniper Tea

Crush an ounce of bruised berries in a quart of boiling water, and steep for five minutes.

Juniper Marinade

The *Joy of Cooking* says a half-teaspoonful of the berries soaked in a marinade or cooked in a stew yields a seasoning that is equivalent to a quarter cup of gin. I make a marinade of crushed garlic, coarse ground pepper, mashed juniper berries, olive oil, and lemon juice for venison. Add the venison and the marinade to a plastic bag, seal tightly, refrigerate, and turn every few hours for 24 hours.

Eating any juniper's berries is said to cure rheumatism. Sufferers are also directed to steep the needles, wood, bark, berries, or all of the above in water, then bathe in the water.

A poultice of mashed juniper berries is said to be antibiotic and soothing to burns. One source claims that eating the berries imparts the smell of violets to the urine (for whatever benefit one might imagine that to be).

RECENT UPDATE

*R*ecent pharmaceutical studies confirm that juniper can stimulate and increase intestinal movements and uterine contractions, affirming its folk use as an abortant. Oil and berries from J. communis were officially listed in the United States Pharmacopoeia from 1820 to 1955 as diuretics.

DESCRIPTION AND LOCALITY

Bayberry (*Myrica cerifera, M. pennsylvanica* or *M. californica*) is a group of evergreen shrubs native to different sections of the country, all put to the same uses. *M. cerifera* is a small southern shrub that occurs in coastal areas and dunes. Bays have gray, stiff, waxy branches and stiff, erect, narrow, and dry aromatic leaves. Their terminal-fruiting branches produce crowded, grayish-green, waxy, aromatic berries. A folk name is **waxberry**. New England bayberry is a tall, semideciduous shrub, but is otherwise the same. *M. californica* is dwarfed and mostly deciduous. All parts of the plant are used. The root bark is taken by heating the roots gently and stripping off the bark. The berries (often called **spiceberries**), leaves, and twigs can be gathered from fall into winter.

EDIBILITY AND RECIPES

Bayberry leaves are a fine seasoning for soups or stews, but its perfume more commonly lends itself to candles and potpourris. Because of the fatty composition of the berries, they are a valuable food for birds and other wildlife in the winter months, but are not palatable to humans.

ETYMOLOGY, HISTORY, AND FOLKLORE

*T*he generic name *Myrica* is from the Greek *myrizein*, to perfume. *Cerifera* means wax-bearing, *Cera* being Latin for wax. Bay's folk names of **candleberry, wax berry,** and **wax myrtle** point to its common use in making candles.

In this country, bayberry was first used in making fragrant candles, possibly the only sweet-smelling thing to be found in a colonial home without refrigeration, running water, or a habit of bathing. Bayberry leaves were handy replacements for Turkish or tropical bay leaves in cooking. The dried berry was ground up and used for snuff, and even today, tobacco snuff is often flavored with aromatics like bay and wintergreen.

REPUTED MEDICAL VIRTUES

*T*he Indians made many medical uses of bay-berry, and the colonists followed suit, finding in the plant a valuable tonic, a powerful astringent, and a stimulant. A mouthwash decoction of bayberry leaf was said to help the spongy gums symptomatic of scurvy, as well as canker sores in the mouth. Drinking the leaf tea helped cure scurvy because of the vitamin C in the

Bayberry Wax for Candles

Boil the berries in water for several hours until the wax floats to the top. Chill, and skim the wax off. The berries may be profitably boiled up to four times before the wax is boiled out.

leaves. Indians used a strong bark tea in hot compresses on varicose veins, while drinking the tea three times a day to cure the varicosity internally. Certain tribes advised women with vaginal discharges to douche daily with bayberry bark tea. Leaf and stem tea was employed against fevers, intestinal worms, liver and kidney problems, as an abortant, and as an insecticide. Among the folk of New England, bayberry bark tea was used to staunch uterine hemmorhage, reduce jaundice, stop dysentery, and clean out the system, among other applications. Finally, certain Amerindians used bayberry as a charm against unfriendly spirits and epidemics.

RECENT UPDATE

*A*ccording to recent analysis, bayberry root bark and plant bark contain several powerful substances, including myricadiol, which influences sodium and potassium metabolism in the body. It is indeed toxic to bacteria, paramecia, worms, and sperm. But it contains tannin, a carcinogen. Therefore, it is probably best used only for candles, or to create a subtle, grayish-green dye.

Mayapples

DESCRIPTION AND LOCALITY

*M*ayapple's single species, *Podophyllum peltatum*, grows all over the eastern half of the United States. It is an erect perennial that grows to about 16 inches, with two parasol-like, deeply-lobed leaves at the top of the plant. In spring it produces a single white flower in the fork of the stem; three months later it yields the large berry that goes by such names as **mayapple, raccoonberry, yellowberry**, and **wild lemon**. The plant is found in dense woods, and the berry ripens in mid-summer. The leaves are distinguishable from other plants by their faded, straw color.

EDIBILITY

Mayapple is highly toxic, with the exception of the ripe yellow berries, which taste like mango, Concord grape, and papaya combined. (A friend once described mayapple as tasting like Kool-Aid. "What flavor?" I asked. "Any flavor," she said.) One of its folk names is **custard apple**, for the ripe flesh has also been described as tasting like egg custard. It can be hard to find the ripe fruit before it has been eaten by woodland animals.

ETYMOLOGY, HISTORY, AND FOLKLORE

Podophyllum peltatum means "duckfooted-leaf-with-shield shape." Mayapple is common and may be the most dangerous of all the plants in Appalachia, and at the same time, the most useful.

REPUTED MEDICAL VIRTUES

Mayapple root was used by Indians, and then by the white settlers, as a cathartic and an abortant, and even (among the Senecas) as a suicide method. The toxin is absorbed through the skin, so even handling the rhizome and roots can be perilous. Mayapple's resin, podophyllin, has been used traditionally against tumors, cold sores, and polyps. Cherokees used it as a vermifuge and a remedy for deafness, and Penobscot Indians removed warts with its resin. Some aboriginal Americans knew that mayapple repelled potato beetles, a claim validated by the U.S. Department of Agriculture in 1984. Native Americans treated venereal warts with mayapple. Testing this old remedy, researchers have found a mayapple derivative that has shown some effect against herpes I and II, influenza A, and measles. Yet it is important to remember that mayapple is a deadly poison.

RECENT UPDATE

A semi-synthetic derivative of mayapple, etoposide, has been used successfully, after conventional chemotherapy has failed, in treating testicular cancer—a disease which is, after automobile accidents, suicides, and AIDS, the chief killer of young men in the United States.

Hawthorn

Hawthorn: spines long as my little finger
that glint in the sun before the leaves come
out, small white flowers like the wild rose
and fruits people don't eat. Virginity.
—*Marge Piercy*

DESCRIPTION AND LOCALITY

Hawthorns (various species of the *Crataegus* genus) are long-thorned, shrubby members of the *Rosaceae* family, with crooked, thorny branches and serrate, often lobed, alternate leaves. The "thorns," which are really just short, sharp branchlets, are as long as six inches. Crataegus is a large genus of "great taxonomic difficulty," according to *Gray's Botany*, which lists 102 known species—most named for their discoverers. Haws are distinctive in that calyx remnants can be seen at the end of each fruit (also the case with rosehips). They grow in small clusters and are terminal.

Hawthorns tend to grow in damp places in the mountains, though certain species thrive anywhere. The fragrant, five-petalled flowers look like sparse roses and are as varied in color. The fruit, known also as **hawthorn berries**, is actually a pome with bony nutlets. Some are black, some bloomy, and some spotted with dark dots.

EDIBILITY

The haws are all safe to eat, though many are unpalatable. Indians ate them and used them in pemmican. The leaves and haws differ enormously within the genus, but in most cases it doesn't matter because the flesh of the fruit is often thin, dry, and tasteless. The tastiest, according to some, are the yellow-fruited hawthorns. A red-fruited European species, *Crataegus azarolus*, grows in Italy, where its fruit is popular for jams and tarts. In the South, some *Crataegus* species, specifically *C. aestivalis*, with reddish spotted haws, are used for **mayhaw** jelly.

ETYMOLOGY, HISTORY, AND FOLKLORE

*T*he genus name comes from the Greek *kratos*, meaning strength. The wood of the plants is tough and brittle, and is used to make durable walking sticks.

REPUTED MEDICAL VIRTUES

*D*ioscorides, writing in the first century after Christ, hailed the medicinal value of the hawthorn for heart ailments in his *De Materia Medica*. In this hemisphere, many Indian tribes also believed eating hawthorn fruit would strengthen the heart. John Brickell, in *The Natural History of North-Carolina* (1737), comments that "the Leaves, Flowers, and Haws, are very binding, therefore good to stop all kinds of Fluxes; the Powder of the Stone drank in Rhenish Wine, is of very great service in the Stone, Gravel, and Dropsie." The bark of the black haw was used as a uterine tonic, according to *American Home Physician* (1857), and a decoction of the bark was a popular remedy in uterine hemorrhage. Dr. Jethro Claus, an 18th-century physician and missionary among the Mohawk

Indians, noted that black haw is astringent, tonic, nervine, uterine sedative, and diuretic. His *Back to Eden* is a classic American herbal.

Catawba Indians beat the bark of the black haw to a mash, soaked it in water, and gave it for dysentery. Haws are used in France today for kidney disease, and in England as a heart tonic.

In 1971, Rodale Press published a book entitled *The Hawthorn Berry for the Heart*. Based on contemporary research, it claims that all parts of the hawthorn contain active principles called flavonoids that act upon the central nervous system in a sedative manner, dilate the blood vessels, and thus lower blood pressure. Because there has not been enough research done to indicate proper dosage, it seems unwise to self-medicate with hawthorn in such crucial areas as the heart and circulatory system. Yet herbalists sell hawthorn tonic, and it is widely believed to reduce the edema (water-retention) caused by heart failure.

RECENT UPDATE

*S*o tough is the wood of the *Crataegus* species that hawthorns have in recent years been planted in the median strips of interstate highways to stop speeding cars from swerving across the divide into oncoming traffic. The spines would undoubtedly wreak havoc with a wax job.

Rosehips

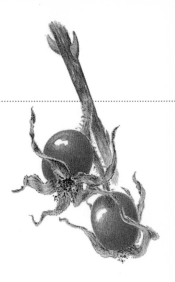

What are all the oranges imported into England
to the hips and haws in her hedges?
—*Henry David Thoreau*

DESCRIPTION AND LOCALITY

*R*osehips, the fruits of rose bushes, are actually *hypanthia,* specialized floral parts, but they look like berries, and are called **roseberries** by some. Like the roses from which they develop, they are borne terminally on the stems of the shrubs. Roses, as most know, have three to five shiny, dark green, serrate, opposite leaflets and thorny stems.

EDIBILITY AND RECIPES

*T*he hips, also sometimes called haws, vary greatly in size and color, but the bright, berry-like fruits are easy to spot on a tramp across freeze-dried, monotone autumn fields, and all are safe to use. They are usually orange to bright scarlet and have a high pectin content, which causes them to make nice sunset-colored jellies. All roses have hips; two particularly popular hip roses are *Rosa rugosa* and *R. eglanteria,* or **sweetbrier.** *Rosa multiflora,* often planted as a "bumper" plant along curves on country roads and along fence rows, has many small, bright red hips that last all winter. Rosehips have a high vitamin C content—about 40 times as much as oranges. About three rosehips will meet a day's requirement of ascorbic acid.

Rosehip Syrup

Measure cleaned hips in a cup or jar. Add an equal amount of sugar and 1/2 the amount of water. Cook slowly until the hips mash easily. Mash them, and press through a strainer. This will keep in the refrigerator, and it's a delicious way to get vitamin C.

ETYMOLOGY, HISTORY, AND FOLKLORE

*A*ctually, "hips" is a nickname of the *hypanthia,* or enlarged *tora,* a *torus* being the flower receptacle below the *calyx.*

REPUTED MEDICAL VIRTUES

*R*osehips appear in all sorts of herbal medicines and teas, and are touted for their antiscorbutic quality. In the lean years of World War II, when citrus fruits were unobtainable, most of northern Europe and Great Britain ate rosehips as an acceptable substitute for fresh fruits and vegetables. The darker the rose, the more flavorful the hip, according to one old saw.

Rosehip Jelly

4 cups rosehips
6 green apples, chopped coarsely
1 lemon
6 cloves
10 allspice berries
1/2 cinnamon stick
2 cups water
1 box dried pectin
4 cups sugar
Cook everything but the pectin and sugar until tender. Mash it all up a little. Cool, and put the mixture through several layers of cheesecloth. If the liquid does not measure four cups, pour a little water through the dregs to make up four cups. Mix liquid and pectin and bring to a boil, stirring all the while. When it boils vigorously, add the sugar all at once and, stirring constantly, bring it to a boil again for 90 seconds. Remove from stove, skim, and jar.

Sumac berries

DESCRIPTION AND LOCALITY

*S*taghorn sumac, *Rhus typhina*, is a shrub that grows up to 15 feet tall. The branchlets are velvety; the leaves are compound, and one to three feet long; each leaflet, four to six inches long and narrow, is toothed. The berries, actually drupes, are roundish, flattened, hairy, and clustered upright like upside-down cones. They are generally red, so they stand out against autumnal or wintry landscapes. **Smooth sumac**, *R. glabra*, is smaller, with fewer hairy berries and branches. There is the white-berried poison sumac, indeed poisonous, but the red-berried ones are useful plants. They grow in depleted land everywhere, at the edges of roads, and in fields. Sumac was an early source of tannin used in leathercraft, and its seeds make a deep black dye.

EDIBILITY

*B*oth sumacs (also spelled **sumach**) are known as **lemonade berry** and **vinegar berry** by virtue of the refreshing drink known as "pseudo-lemonade."

Pseudo-Lemonade

Soak a stalk of the tightly packed red berries overnight in a pitcher of cold water. The berries contain malic acid, as does vinegar. If necessary, strain through cheese-cloth to remove the hairs.

REPUTED MEDICAL VIRTUES

*I*ndians dried the fruit heads for winter use, and made poultices of the astringent berries to stop bleeding. Hemorrhoids were treated with sumac berries and roots pounded to a mash and applied as poultices, as were warts, fever blisters, and canker sores. John Brickell noted that sumac leaves and seeds "outwardly resist putrefaction, drie up running Sores, heal old Ulcers, Gangrens, &tc." Sumac was also considered a vermifuge by some Indians.

A tea made from the berries was believed to loosen mucous accumulations in the throat. Decoctions of root bark and plant bark have been used in the treatment of venereal diseases, diarrhea, and other internal complaints, various sorts of dermatitis, and diabetes. In Appalachia, sumac is widely believed to lessen the severity of asthma, to cure dysentery, and to soothe sore throats. Sumac leaves have traditionally been rolled and smoked to stop asthma attacks. In fact, sumac has been tried as a tobacco substitute and is said to be useful in curing a tobacco habit.

Saw palmetto

DESCRIPTION AND LOCALITY

Saw palmetto (*Serenoa serrulata*) berries grow on a dwarf palm in coastal areas in the extreme southern part of the United States. The flat, yellow-green leaves grow in a fan shape, almost in a circle, one to three feet in diameter. The berries, called **serenoa**, and **sawpalm** as well as saw palmetto, are edible, black oval drupes that ripen in late fall.

ETYMOLOGY, HISTORY, AND FOLKLORE

One retailer of herbs in Boone, North Carolina, says that saw palmetto berries sell more briskly than ginseng. When I asked why, he would only shrug. The name in Latin means clear or bright, kin to "serene" in English.

REPUTED MEDICAL VIRTUES

The Mayans used the pulp as a remedy for snakebite and insect bites, and poulticed ulcers with it. Early Americans used the sweet pulp as a purgative. They made brooms from the branches, while in the Caribbean, hats and baskets were woven from the flat leaves. From 1906 to 1916, palmetto berries were listed in the *United States Pharmacopoeia* as diuretic, sedative, and anticatarrhal.

I began to understand the saw palmetto's popularity when I learned that Southern settlers saw animals and Indians grow fat on the oily berries, and ascribed vitalizing properties to them. Today, certain folk in Appalachia believe they are even more powerful—that saw palmetto berries can cure enlarged prostates (which afflict most men after middle age), increase breast size, pep up sexual vigor in both genders, and increase sperm production. They are even supposed to be able to reverse atrophy of breasts, testes, and female genitalia! No wonder the berry is a best-seller.

Wintergreen

DESCRIPTION AND LOCALITY

Wintergreen (*Gaultheria procumbens*), also called **teaberry, checkerberry, checkberry, boxberry, redberry, groundberry,** and **roxberry**, is indigenous to the eastern part of America, and is found in temperate-climate woods.

Teaberry most often grows in dense, dry evergreen woods, and lives in symbiosis with pine trees. The tiny, neat plant is usually no more than six inches high, and the eye can spot in wintry woods the bright red, globe-shaped berry and the dark green, glossy, ovoid, one- to two-inch leaves. In the summer, before the berries ripen, the plant has tiny, nodding, bell-like, white flowers in the leaf axils. The berries are edible, even pleasant when ripe, but it is hard to believe that they were once sold as a fruit. They hang, keeping their bright color all winter long, then rot or drop to the ground, releasing their seeds to form new plants. They are an important article in the diet of at least three woods creatures in the lean winter months, hence three of their common names: **partridgeberry, grouseberry**, and **deerberry**.

EDIBILITY

We associate teaberry or wintergreen flavor with products such as mouthwash, chewing gum, candies, toothpaste, and even with fresh-tasting snuff. It is still a favored seasoning for cough drops or sore throat remedies. Teaberry oil is toxic, containing salicylate. One-quarter teaspoon contains as much as half a dozen aspirin, and thus might be fatal to a child. It's okay for linament, but don't use it as a seasoning.

HISTORY AND FOLKLORE

In 18th-century New England, following the disposition of an entire shipload of tea into Boston Harbor, many substitutes were tried. Most were found to lack the *je ne sais quoi* of a good cup of tea. One of the least offensive, however, was wintergreen leaf tea. Teaberry was supposed to increase the flow of urine (though perhaps that was only the result of drinking many cups of the fragrant, slightly biting tea).

REPUTED MEDICAL VIRTUES

*W*intergreen tea was taken for rheumatism, fevers, and lumbago by scores of different Indian tribes. Mountain folk still believe that drinking wintergreen leaf tea, hot and pungent, relieves many kinds of pain—specifically rheumatism and arthritis pain. Wintergreen oil is a counter-irritant, making it popular as an ingredient in rubs for muscular aches and pains. A counter-irritant plus the rubbing causes heat, increasing blood flow to the area of the body being treated, which may help wash away products from the tiny rips in muscles that occur with over-exertion. Thus it eases both swelling and discomfort, and hastens healing. But there may be another explanation: One pain can be masked by another, so the burning sensation may simply override the ache. Teaberry leaf compresses have long been used to reduce swelling of tissues around inflamed joints, though the fragrance today is usually artificially formulated or made from the bark and twigs of young black birch trees.

RECENT UPDATE

*T*oday it is known that teaberry contains arbutin, the same substance that, in cranberries and blueberries, helps end the misery of chronic bladder infection. Teaberry oil, methyl salicylate, is close kin to aspirin.

Physicians speculate, though, that athletes respond to wintergreen-scented products because they trigger memories of happy times spent with teammates in locker rooms smelling of linament. Thus, a sort of placebo effect is suspected.

Pokeberries

DESCRIPTION AND LOCALITY

*P*oke is a perennial weed of the temperate zone that grows to a man's height or higher, with bright green leaves, a red stem, white to pinkish flower, and long, loose clusters (racemes) of shiny, dark purple berries that stay into the winter. It grows in the eastern part of the United States, some parts of the western United States, and even in Hawaii. It is a striking plant in any landscape.

EDIBILITY AND RECIPES

I have observed birds eating **pokeberries**. I presume the birds did not fly off and die. There is a lesson here: The fact that a bird or another animal eats a plant does not mean it is safe for human consumption. Pokeberry (*Phytolacca americana*), also known as **poke, skoke, garget**, and **pigeonberry**, is highly toxic, containing a saponin mixture called phytolaccatoxin. However, the young tender shoots of early spring are an exception, and are often eaten as a green in the southern mountains.

ETYMOLOGY, HISTORY, AND FOLKLORE

*P*oke is an Algonquin name. Its scientific name comes from the Greek *phyton*, or plant, and *lacca*, or red. Poke has long been used to dye cloth orange, purple, brown, or red. The berries, blackish-purple, make an ink that some say outlasts modern commercial inks—hence another folk name, **inkberry**. Pokeberry reportedly yields a true red dye.

REPUTED MEDICAL VIRTUES

*S*everal years ago, a man in one of my Elderhostel classes told me, as we were discussing folk medicine in Appalachia, that his Missouri-born grandfather had been famous in several states during the mid-19th century as an arthritis doctor. He successfully treated folks with his own personal medicine, allegedly learned from Indians, the chief ingredients of which were whiskey and pokeberries.

John Josselyn first mentioned poke in America, saying the Indians "cure their wounds with it." Poke may hold the record for having more reputed curative powers than any other plant in this country. Indians also used pokeberries to treat syphilis, scrofula, and skin diseases. They gathered the roots of poke, pounded them into a poultice, and applied them to cancers, swellings, ulcers, both human and bovine mastitis, and old scabs. Bovine mastitis, by the way, is garget, one of the folk names of the plant. Pamunkey Indians boiled the berries for a rheumatism medicine.

A mixture of poke root-pulp and lard has been used with some effect on eczema and psoriasis. In colonial times, pokeberries were first adopted by the Englishmen, then exported to England where they achieved popularity.

Cadwallader Colden, an 18th-century physician and botanist, wrote a History of the Five Indian Nations, which included the use of pokeroot as a cancer cure. So impressed was Colden that he passed the information on to Benjamin Franklin, adding that pokeroot would also rid one of corns overnight. As recently as June of 1993, a Virginia newspaper article claimed that poke is a cure for baldness, round worms, obesity, wrinkles, gray hair, hot flashes, bad dreams, black lung disease, and ingrown toenails.

Pokeshoots

Cut young shoots at ground level when no more than seven inches high. Parboil, drain, then cook until tender in new water. They are tasty and useful, and they contain a fair amount of vitamin C.

Fried Poke

Flour up the young stalks and fry them in bacon fat. They are said to taste just like asparagus. Or flour up the young leaves and fry them in bacon fat. These are reputed to taste like fish.

Pokesallet

Take a mess of poke shoots, wash it, and chop it coarsely. Boil for 30 minutes. Fry some fatback in a skillet while it is boiling. Drain the poke, salt it, and fry it in the leftover grease. Serve the fried poke with the fried pork and some cornbread. Sprinkle cider vinegar on the greens and drink buttermilk with it if you want the whole pokesallet experience.

*P*okeberries have a principle that inhibits *Herpes simplex*, and is generally effective against non-retro-viruses. It is now certain that poke does contain a substance that inhibits cell division. Recently it has been discovered that pokeberries have a powerful anti-HIV agent called pokeweed antiviral protein (PAP). How exciting it would be if this common plant could provide us with a cancer cure, or a rescue from the scourge of AIDS.

Actaea rubra

Hollyberry

BERRIES TO AVOID

*O*ther familiar berries that are toxic include **hollyberries** (even though birds eat them safely, people cannot); **buckthorn** (*Rhamnus cathartica* and other species); and **twinberries** (*Lonicera vars*). **China berries**, **baneberries** (*Actaea rubra*), **privet berries**, **mistletoe berries**, **bittersweet berries** (both *Celastrus scandens* and *Solanum dulcamara*), **ivy berries**, **pyracantha berries**, **pigeon berries** (*Duranta repens*), berries of the **vervain** family, **ornamental nightshade berries** (though some report that these are edible when cooked), **blue cohosh berries**, **poison ivy berries**, and **asparagus berries**.

Avoid them, one and all.

Left: Celastrus scandens

Solanum dulcamara

Minor Berries

. . . a kind of paste or pudding, made of the flour of the "pomme blanche," as the French call it, a delicious turnip of the prairie, finely flavored with the buffalo berries which are collected in great quantities in this country, and used with divers dishes in cooking, as we in civilized countries use dried currants, which they very much resemble.

—from the journal of the artist George Catlin (1830)

Bearberries

A northern plant, *Arctostaphylos Uva-ursi*, the **bearberry** is also called **mountain cranberry**, **arberry**, **mealberry**, **Uva-ursi**, and **hog-cranberry**. Bearberries grow in evergreen forests all over Canada and throughout the northern half of the United States. Bearberry is a thick, matted, evergreen shrub with stems six inches high, reddish bark, bright green leaves, and long, white to pink flowers in clusters. Its main root sends out several underground stems, from which grow branching stems. It appears in gardens as a groundcover.

Bearberry is unimportant as a food for humans; it is a reddish, mealy, dry drupe one to two inches in diameter, obviously eaten by bears, and it is an important wildlife food in its native areas. *A. Uva-ursi* enjoys success in modern folk medicine as a key ingredient in kidney or bladder teas, marketed widely in Europe and the Orient. Though the plant in this country is identical to the one in Europe, Europeans much prefer the American bearberry. Americans, of course, prefer the European bearberry.

Buffalo berries

*B*uffalo berries (*Shepherdia argentea,*) also called rabbit berries, are a western species, growing from Minnesota to Kansas and westward. A tall, deciduous shrub, buffalo berry grows in moist valleys and low meadows, and is noticeable for its oblong, opposite silver leaves, its rusty, dull whitish branches, and its small, thorny branchlets. The fruits, roundish and dull red, are drupes growing in axillary clusters and ripening in late summer or early autumn. The seed is elongated and has ridges. The flesh is acidic and pleasant, slightly astringent.

Buffalo berries have traditionally provided Indians, buffaloes, and rabbits with a pleasant, healthful fruit and a red dye. Some writers claim the berries are too sour to eat out of hand, but like persimmons and certain grapes, they are tamed by the first frost into sweetness. Indians in the Midwest made a buffalo berry sauce similar to cranberry sauce. The berries were also dried and eaten like raisins.

Buffalo berries have recently become naturalized outside their native habitat through their inclusion in eastern gardens. In the 19th century, unsuccessful efforts were made to market the fruit commercially.

A similar species, *Shepherdia canadensis*, the **Canadian buffalo berry**, has yellowish-green blossom clusters, egg-shaped leaves a little shorter, and slightly elongated berries that are more orangy in color and of questionable edibility. *Gray's Botany* refers to them as "nauseous."

Scarcely distinguishable from buffalo berries are **silverberries** (*Elaeagnus commutata*). In this species, the leaves are alternate, and there are no thorny branches. The berry, really a drupe, is edible but not desirable, being silvery, dry, and mealy. Similar is the **Oleaster**, or **Russian olive** (*E. angustifolia*) with narrower leaves, the same silvery branches, and sweet, edible drupes. Both grow not only in the West and Midwest, but also in many places in the East, where they have escaped from cultivation.

Bunchberries

Bunchberry (*Cornus canadensis*) is an herbaceous, low-growing (three to six inches high) member of the dogwood pack. Its cyclic leaves grow around a stem, with an umbel of red, edible berries that are sometimes called **crackerberry**, **crackberry**, and **pudding berry**. The flower clusters are four-petaled and white, and the herb rises from tough, slender rhizomes.

I can find no uses for the bunchberry, which is said to be "insipid." In fact, its edibility is questioned by many, while others find it good. Because many related forms have been noted, it is possible that some have berries worth gathering. A similar species, *C. suecia*, which grows in Canada and Scandinavia, is used by Laplanders to make a pudding of whey and crushed *Cornus* berries. Perhaps the American version, a look-alike, earned its reputation on the back of its better-tasting relative.

Cape Gooseberry

$\mathscr{C}$ape gooseberries are mostly subtropical or tropical plants. They bear edible berries that are generally yellow to orange to red. Cape gooseberries were introduced to Africa from Peru, and they are now commercially produced on the Cape of Good Hope in South Africa. The name of the genus, *Physalis,* means bladder, and the berries are enclosed in a bladder-like calyx. No kin to gooseberries, they are, of all things, tomatoes. The cape gooseberry, which looks naturally rather like a gooseberry, is also called **dwarf cape gooseberry, tomatillo, jamberry, ground cherry, husk tomato,** and **goldenberry,** among other names. It is also called, just to

Crowberries

Crowberries, also known as **curlew berries** (*Empetrum nigrum*), an arctic plant, are a common evergreen with short needles alternate or cyclic, which grows across the northern part of this continent to the southern reaches of the Great Lakes. Crowberries grow on low, matted, spreading shrubs, the fruits blackish (crow-colored) with a hard seed. The procumbent plant is an important wildlife food, with fruits available year-round. The purple-black berries have a medicinal taste, which lessens after they have been frozen or cooked.

Crowberries are said to make an excellent jelly and a good beverage. The jelly is made with two cups of water and three cups of berries cooked together and strained, one box of pectin, and sugar—one cup for each cup of the resulting juice. Follow the simple directions in the pectin box for an interesting and unusual jelly.

A similar species, *Empetrum Eamesii* (named for botanist Edwin Eames), is **rockberry**. The branches are whitish, and the small berries are light red, with thin translucent skin and white flesh.

Hackberries

ackberries are indigenous to Canada and the United States as far south as Tennessee. They are found in rocky woods along creeks and sometimes along roads. The hackberry tree can be big, with a straight trunk and leaves that look like the leaves of their kin, the elms. Hackberries themselves are small orange, red, yellow, or black drupes the size of blueberries. They are sweet and edible, and are borne singly and terminally, looking somewhat like a wild cherry.

Hackberry, *Celtis occidentalis*, goes by many nicknames: **hagberry, bird cherry**, and **sugarberry** are but three. They are anciently identified with humans, for the Choukoutien site where "Peking Man" was found revealed so many broken shells of hackberry seeds that it seems likely they had been ground into paste or flour. The genus name *Celtis* is from the ancient Greek word for the Lotus-berry, upon which the Lotus-eaters of the *Odyssey* dined.

American Indians used hackberries as food, including pemmican. In addition, certain Indians treated syphilis with a decoction of the fresh inner bark of hackberry, which is astringent. Some western Indians dried and pulverized the fruit into a kind of flour. As with other astringent plants, a tea made from hackberry bark was believed to be a tonic.

142

Partridgeberries

A plant often confused with wintergreen is *Mitchella repens*, the **partridgeberry**. The partridgeberry is a perennial evergreen trailing vine with small, shiny, and roundish opposing leaves. The bright red, paired, 1/4-inch berries remain on the plant all winter; hence the nicknames **scarlet**

berry and **twinberry**. Most sources list partridgeberry as edible, but some list it as poisonous. As it is eaten by deer and partridge (like wintergreen), it is also called **deerberry**. In bloom it would not be confused with wintergreen, as its flowers are paired and white with a rosy or purplish tint.

Partridgeberry is in fact edible, slightly aromatic, and not unpleasant, though some Wisconsin Indians called it **stinkberry**. A tea made in Appalachia of the entire plant, harvested in the fall, is said to aid childbirth, alleviate diarrhea, and stop nightmares. It is also known as **squawberry**. It contains tannin, an astringent now considered a carcinogen, and it was believed to staunch excessive menstrual flow. It was also considered diuretic by Indians.

The answer to why the plant is designated poisonous may lie in the fact that another plant, *Duranta repens*, or **pigeonberry**, a large tropical shrub with white flowers and masses of deadly orange berries, is also sometimes called partridgeberry. *D. repens*, however, grows no farther north than the Caribbean, Mexico, and southern Texas.

Snowberries

Snowberry (Symphoricarpos albus or *S. racemosus)* is a deciduous, shrubby member of the honeysuckle family about three to six feet tall. It has simple, opposite, oval, two-inch leaves edged with broad, wavy, irregular teeth, and tiny, pink, bell-like flowers in close, short spikes or clusters. Hence the genus name, which means "bearing fruit together." Snowberry, also known as **ivory plum**, grows in the northern two thirds of the United States, bearing in the autumn a profusion of snow-white, axillary berry clusters that are "toxic if ingested in quantity" but sweet-tasting and not seriously poisonous, according to plant researchers at the Royal Botanic Gardens in Kew, London.

However, the **snowberry** is listed in many sources as a poisonous gastrointestinal irritant. The berries contain chelidionine, a central nervous system depressant alkaloid. Ingestion of too many of the berries could cause vomiting, dizziness, depression, and sedation. However, some Indians made a root tea from the plant that was said to help expel afterbirth. One species, *S. orbiculatus*, bears red berries.

Robert Frost, in what Randall Jarrell once opined might be the most erotic lines ever written in English, mentions the gathering of snowberries. The context of the poem, "The Pauper Witch of Grafton," suggests that they may be used as part of a potion or spell. Here are the lines:

I made him gather me wet snow berries/ On slippery rocks beside a waterfall/ I made him do it for me in the dark/ And he liked everything I made him do.

Creeping snowberry, *Chiogenes hispidula*, is an evergreen creeper. Thoreau wrote that a tea made from its leaves was better than black China tea.

Oddly, a close relative of the snowberry with a more alarming name, *S. occidentalis*, or **wolfberry**, is benign. It is a western plant with a greenish-white fruit that soon darkens to brownish-purple or even black. There seems to be little information about its edibility, so we might conclude that it is not an especially useful fruit. Certain Indian tribes made a wolfberry eyewash from the leaves of the plant.

Closely related to snowberries and wolfberries are **coralberries**. Coralberry (*Symphoricarpos orbiculatus*) produces opposite, entire leaves and clustered berries on the stem, and causes gastrointestinal upsets. Also called Indian berry, the fruit is coral-pink to purple. It is apparently not an important fruit to humans, and no medical uses are attributed to it.

Creeping snowberry

Creeping snowberry

Good Memories
for When
You're Old

On June 28, 1993, I stopped on a well-traveled path leading to a university and a nursery school and, within two feet of the walkway, I picked about a pint of ripe black-caps. I ate my fill of the berries with their mellow sweetness and hauled two big handfuls home. I called the nursery school when I got home to tell the teacher I'd left plenty of berries down low for her pint-sized horde. But she'd have none of it; there might be snakes, and the children might get scratched up.

I think one of the criteria for skillful teaching of young children involves adopting what might be called a life perspective: For the rest of this child's life, how will what I teach him serve him? How will this lesson enhance her days for the next 50 or 80 years? Viewed in this way, the gathering and eating of wild berries might be a great deal more useful in the long haul than skill at soccer or football.

∞

Having thus far noted remedies real and imaginary for most conceivable human ills, I now offer my own personal folk remedy. I would like most seriously to suggest berries as a remedy for depression.

Berries are a free and delicious low-calorie treat crammed with vitamin C, potassium, magnesium, fiber, and fructose. Hunting and gathering them will increase circulation, slow the heartbeat, lower the blood pressure, clear out the head, lungs, and heart, and slow the pace of living to a memorable, reasonable rate. Wild berries are organically grown, too.

Since jogging has become a popular sport, wild asparagus, which tends to grow alongside country roads, has become scarce where it used to be abundant—perhaps because runners can grab it without breaking aerobic stride. But wild berries must be stopped for, and brambles scratch unprotected legs, so the berries are passed by.

We all move too fast these days—to catch the subway or to cut in line before the light changes. Instead, while we wait for the light we might look for a sandy strip along an old railroad track where the sandy soil and full sun have for a century and a half invited wild strawberries to grow. We might decide to spend the day picking these tiny fruits, taking them home, and making them into sublime shortcakes to end the day or tiny jars of jelly to delight us in winter with their summer perfume.

We breathe polluted air outside and conditioned air inside. We might try the respite of a huckleberry patch in August with nobody around but us, with the birdsongs, the breeze, the bees, and the reward of an afternoon's picking: huckleberry cake with hard sauce for supper.

We too infrequently these days spend time pleasantly in conversation with friends, engaged in mutual enterprise. What pleasure in meeting on a morning after a hard frost, our breath making clouds in front of us, heading out for the morning to gather rosehips bright against gray fields!

Back in the late forties, when the phenomena of therapy and psychoanalysis began to become common, my grandfather said, musingly, "If you have a good friend of the same sex, I can't think why you'd ever have to go be analyzed. Isn't that what friends are for?" Simplistic today, of course, but not without merit. If you don't have such a friend, it is never too late to make one. Select someone interesting, invite him or her to gather blackberries with you, and see what happens. If we did more things of this sort, we'd likely be healthier in body and in mind.

Many years ago, when I was thinking of marrying my husband, one of the tests I devised was a berry-picking expedition. I was a veteran by then of several disastrous sallies towards matrimony, one involving a fellow who (during his wild-food trial) asked me to show him a pickle plant. Adding insult to injury, he refused to believe me when I showed him cucumbers growing in my own back yard. Another suitor got such a terrible case of poison ivy that he never came near me or Virginia again. But the man I married did fine, picking blackberries fearlessly all afternoon and asking to do it again. We still pick wineberries along a certain road, something we've done for all 30 years of our marriage and child-raising.

And when winter comes and there are no more berries, the memories will be there, concentrated in jelly and jam or ready to be taken out of the freezer to remind us that summer is not just a dream.

I ask you, as a friend once asked me: "Don't you want to have good memories for when you're old?"

Bibliography

Bailey, L.H. *Cyclopedia of American Agriculture*. New York, Macmillan, 1907.

Bailey, L.H., and Bailey, Ethel Zoe. *Hortus Second*. New York, Macmillan, 1930.

Berglund and Bolsby. *Edible Wild Plants*. New York, Charles Scribner's Sons, 1977.

Booth, Sally Smith. *Hung, Strung, and Potted*. New York, Clarkson Potter, 1971.

Brickell, John. *The Natural History of North-Carolina, With an Account of the Trade, Manners, and Customs of the Christian and Indian Inhabitants*. Dublin, James Carson, 1737.

Brittan, Nathaniel L., and Addison Brown. *Illustrated Flora of the Northern U.S. and Canada*. New York, Dover, 1971.

Brothwell, Don and Patricia. *Food in Antiquity*. New York, Frederick A. Praeger, 1969.

Brown, Susan Anna. *The Book of Forty Puddings*. New York, Charles Scribner's Sons, 1882.

Brunvand, Jan Harold. *The Study of American Folklore*, Second Edition. New York, W.W. Norton and Company, 1978.

Coon, Nelson. *Dictionary of Useful Plants*. Emmaus, Pa., Rodale, 1974.

———. *Using Wayside Plants*. New York, Hearthside Press, Inc. 1957.

Coxe, William. *A View of the Cultivation of Fruit Trees with Management of Orchards & Cider, &c.* 18 Vols. Philadelphia, J. B. Lippincott, 1817.

Culpeper, Nicholas. *Culpeper's English Physician and Complete Herbal*. London, Nathaniel Brook, 1652; the British Directory Office, 1798.

Daly, Douglas. "Tree of Life." *Audubon*, March/April 1992.

Darrow, George M. *The Strawberry*. New York, Holt, Rhinehart, and Winston, 1966.

Doughty, J.V.T. *Cabinet of Natural History and American Rural Sports*. Philadelphia, J.B. Lippincott, 1830. (illustrations)

Downing, Alexander Jackson. *Downing's Fruits and Fruit Trees of America*. New York, John Wiley, 1845.

Eastwood, B. *Complete Manual for the Cultivation of the Cranberry*. New York, Orange Judd and Company, 1866.

Embury, Emma C. *American Wild Flowers in their Native Haunts*. New York, D. Appleton and Company, MDCCCXLV (1845) (illustrations)

Everett, T.H., ed. *New Illustrated Encyclopedia of Gardening* unabridged. New York, Greystone Press, n.d.

Fernald, Merritt L. *Gray's Manual of Botany, 8th Edition*. New York, American Book Company, 1966. (First Edition 1848.)

Fielder, Mildred. *Plant Medicine and Folklore*. New York, Winchester Press, 1975.

Figuier, Louis. *La Terre Et Les Mers*. Paris, Librairie Hachette, 1892.

Fischer, David Hackett. *Albion's Seed*. New York, Oxford University Press, 1989.

Freeman, Margaret B. *Herbs for the Mediaeval Household*. New York, The Metropolitan Museum of Art, MCMXLIII.

Fuller, Andrew S., Editor. *Woodward's Record of Horticulture No. II*. New York, Francis W. Woodward, 1868.

Gerard, John. *The Herball*. London, Flannery and Son, 1597.

Gibbons, Euell. *Stalking the Faraway Places*. New York, David McKay, 1973.

———. *Stalking the Healthful Herbs*. New York, David McKay, 1966.

Grant, Verne. *Plant Speciation.* New York, Columbia University Press, 1971.

Grimm, William Carey. *Recognizing Native Shrubs.* Harrisburg, Pa., Stackpole Books, 1966.

Hanson, Betsy. "Yews in Trouble," *Discover Magazine,* January 1992.

Harland, Marion. *Breakfast, Luncheon, and Tea.* New York, Charles Scribners's Sons, 1875.

Harlow, William M. *Fruit Key and Twig Key to Trees and Shrubs.* New York, Dover, 1941.

Hartley, Dorothy. *Lost Country Life.* New York, Pantheon Books, 1979.

Heinerman, John. *Heinerman's Enclyclopedia of Fruits, Vegetables, and Herbs.* Nyack, New York, Parker Publishing Company, 1988.

Heiser, Charles B. *Of Plants and People.* Norman, University of Oklahoma Press, 1985.

Henderson, Charles. "Cancer Drug, Insecticide, found in Common Tree," *Cancer Weekly,* February 17, 1992.

Henkel, Alice. *Wild Medical Plants of the United States.* Bulletin #89, U.S. Department of Agriculture, 1906.

Hiscox, Gardner D. *Henley's Twentieth Century Book of Recipes, Formulas, and Processes.* New York, Norman W. Henley Publishing Company, 1907.

Hitchcock, Susan Tyler. *Gather Ye Wild Things.* New York, Harper and Row, 1980.

Hobhouse, Henry. *Seeds of Change.* New York, Harper and Row, 1985.

"It's fall—be alert to children eating berries" *Child Health Alert,* October 91.

James, Wilma Roberts. *Know Your Poisonous Plants.* Happy Camp, California, Naturegraph Publications, 1973.

Josselyn, John. *New-Englands Rarities Discovered.* London, 1672. Reprint Boston, Massachusetts Historical Society, 1972.

Kerr, Elizabeth M., and Aderman, Ralph. *Aspects of American English.* Harcourt, Brace & World, Inc. 1963.

Krochmal, Arnold and Krochmal, Connie. *A Guide to Medical Plants of the United States.* New York, Quadrangle, New York Times Book Company, 1973.

Krochmal, Arnold, et al. *A Guide to Medicinal Plants of Appalachia.* Agriculture Handbook No. 400, U.S. Department of Agriculture, 1971.

Lampe, Dr. Kenneth F. and McCann, Mary Ann. *AMA Handbook of Poisonous and Injurious Plants.* Chicago, American Medical Association, 1985.

Landreth, John Claudius, D & G. Nursery and Seedman: *The Floral Magazine and Botanical Repository.* Philadelphia, 1832. (illustrations)

Leighton, Ann. *Early American Gardens.* Amherst, The University of Massachusetts Press, 1986.

Loudon, J.C. *An Encyclopaedia of Gardening, Vols. I and II.* London, Longman, Rees, Orme, Brown, Green, and Longman, 1835. Reprint New York, Garland Publishing, Inc. 1982.

MacKeever, Frank C. *Plants of Nantucket.* Amherst, University of Massachusetts Press, 1968.

Marks, Geoffrey. *The Medical Garden.* New York, Scribner, 1971.

McCaleb, Rob. "Bilberry for Circulation." *Better Nutrition for Today's Living,* July 1992.

McIntosh, Charles. *The Orchard and Fruit Garden.* London, William S. Orr and Company. Amen Corner, 1839.

Meyer, Clarence. *American Folk Medicine.* New York, Thomas Y. Crowell, 1973.

Miller, Amy Bess. *Shaker Herbs, A History and Compendium.* New York, Clarkson N. Potter, 1976.

Millspaugh, Charles F. *American Medicinal Plants.* New York, Dover 1974. (Originally published in 1892.)

Monardes, Nicholas. *Joyfull Newes Out of the New-Found Worlde.* London, E. Allde by assigne of B. Norton, 1596.

Morse, Sidney. *Household Discoveries.* Petersburg, New York, Success Company, 1908.

Muench, Frederick. *School for American Grape Culture.* Conrad Witter, St. Louis, Missouri, 1865. Distributed by J.B. Lippincott, Philadelphia, 1865.

Muenscher, Walter C. *Poisonous Plants of the United States.* New York, Macmillan, 1939.

Nuttall, G. Clarke. *Beautiful Flowering Shrubs.* New York, Frederick A. Stokes Company, 1922.

Nuttall, Thomas. *Introduction to Systematic and Physiological Botany.* London, J. Morpheus, 1830.

Paxton's Magazine of Botany and Register of Flowering Plants. London, Orr and Smith, MDCCCXXXIV (1834).

Pizer, Vernon. *Eat the Grapes Downward.* New York, Dodd, Mead, and Company, 1983.

Pliny the Elder. *Historia Naturalis.* Reprint Cambridge, Harrup and Sons, 1898.

Pliny the Elder. *Medicina Plinii.*

"Pokeweed Antiviral Protein," *Nature,* September 6, 1990.

Pond, Barbara. *A Sampler of Wayside Herbs.* New York, Crown Publishers, 1974.

Root, Waverley, and Richard de Rochemont. *Eating in America.* New York, The Ecco Press, 1981.

Sann, Paul. *Fads, Follies, and Delusions.* New York, Crown, 1967.

Schneider, Elizabeth. "A Berry Primer," *Gourmet,* July 1993.

Simmons, Amelia. *American Cookery.* London, 1796.

Strehlow, Dr. Wighard, and Hertzka, Gottfried, M.D. *Hildegard of Bingen's Medicine.* Santa Fe, New Mexico, Bear and Company, 1988.

Strong, A.B., M.D. *The American Flora, or History of Plants and Wild Flowers.* New York, Green and Spencer, 1845.

Sutton, Ann and Sutton, Myron. *The Audubon Society Nature Guides: Eastern Forests.* New York, Alfred A. Knopf, 1985.

Tannahill, Reay. *The Fine Art of Food.* Norwich, England, The Folio Society Ltd., n.d.

Terwilliger, Karen, ed. *Virginia's Endangered Species.* Blacksburg, Virginia, McDonald and Woodward Publishing Company, 1991.

Thomas, John J. *The American Fruit Culturist.* Auburn, Derby, Miller, and Company. 1851.

Thoreau, Henry David. *Faith in a Seed.* Island Press, Shearwater Books, 1993.

Tierney, John. "A Patented Berry Has Sellers Licking Their Lips" *New York Times,* October 14, 1991.

Torrey, John. *Natural History of New York Botany.* New York, Appleton and Company, 1843. (illustrations)

Tusser, Thomas. *Five Hundred points of Good Husbandry.* London, J. Morphew, 1710.

Tyler, Varro E., Ph.D. *The Honest Herbal.* Philadelphia, George F. Stickley Company, 1981.

Vogel, Virgil J. *American Indian Medicine.* Norman, University of Oklahoma Press, 1970.

Weiss, Gaea and Shandor. *Growing and Using the Healing Herbs.* Emmaus, Pennsylvania, Rodale Press, 1985.

"What Plant is the Most Useful Medicinally?" *International Wildlife,* January-February 1986.

Wheelwright, Edith Grey. *Medicinal Plants and Their History.* New York, Dover, 1974.

Wilcox, Estelle. *New Dixie Cookbook, Revised and Enlarged.* Atlanta, Georgia, Dixie Cook-Book Publishing Co., L.A. Clarkson & Co, 1895.

Wilkinson, Albert E. *Encyclopedia of Fruits, Berries, and Nuts, and How to Grow Them.* Philadelphia, New Home Library, The Blakiston Company, 1945.

Williams, Susan. *Savory Suppers and Fashionable Feasts.* New York, Pantheon, 1985.

Wren, R.C. *Potter's Encyclopedia of Botanical Drugs and Preparation.* London, Pittman, 1956.

Wyman, Donald. *Wyman's Gardening Encyclopedia,* Rev. New York, MacMillan, 1977.

APPENDIX A

Glossary of Health and Medical Terms

alkaloid: a vegetable base, often poisonous

anthocyanin: a vegetable compound with various benefits

antibiotic: bacteria-destroying

anticatarrhal: fights watery discharges of the nose and throat

anti-inflammatory: fights inflammation

antiscorbutic: fights scurvy

antispasmodic: fights spasms (generally of the gut)

aperient: laxative

astringent: causing contraction of tissue

bitters: flavored substances supposed to aid digestion

cathartic: violently laxative

counter-irritant: a substance that masks one pain by causing another

cultivar: a cultivated variety

cystitis: bladder inflammation

decoction: boiled-down liquid

dentifrice: tooth-cleaner

diaphoretic: perspiration-inducing

diuretic: urine-inducing

dropsy: fluid retention

dysentery: inflammation of intestines causing diarrhea

eczema: skin inflammation

edema: water retention

febrifuge: causes fever to flee

flavanoid: a plant principle with medical properties

gout: inflammation of joints due to excess uric acid

herbal: a book about herbs and their properties, generally medical

hydrolyzable: breaking down by a chemical reaction with water

infusion: steeped liquid

lumbago: rheumatism in loins

mastitis: inflammation of milk glands

nervine: nerve-soothing

phytolaccatoxin: the poison of poke

poultice: a soft mass applied to inflamed part

psoriasis: scaly skin

purgative: strongly laxative

rheumatism: joint pain and inflammation

salicylate: the active principle in aspirin

saponin: a plant substance causing foaming

scrofula: a tuberculosis causing glandular swelling

scurvy: a disease caused by lack of Vitamin C

sedative: calming

solanine: a toxin of the potato family

tannin: an astringent substance in tree bark

tincture: an alcoholic solution

tonic: invigorating

toxin, toxic: poison, poisonous

uterine hemmorhage: unusual bleeding from the uterus

vermifuge: causes worms to flee

An Index of Berries

In this list of berries, the

(=) sign means a synonym, and the

(p) is a poison berry. The fruits so marked probably have valuable uses, but the berries are not to be eaten.

(B) indicates that the fruit not called a berry is a true berry.

(D) is a drupe,

(T) is a tomato,

(P) is a pome, and

(A) is an aggregate fruit.

Omitted from this book, and this list, are quite a few true berries (tomatoes, persimmons, and oranges, for instance) on the premise that they are not *known* as berries.

American cranberry see mountain cranberry
American huckleberry see huckleberry
arberry =bearberry, 134
bababerry =*Rubus* hybrid, 49
baked apple berry =cloudberry see also blackberry
balloon berry *Rubus illecebrosu;* = strawberry raspberry, 46
baneberry *Actaea alba, rubra;* =snakeberry =chinaberry (p), 12, 129
barberry *Berberis vulgaris, Mahonia aquifolium;* =berberry =mountain grape =Rocky
  Mountain grape =oregon grape =jaundice berry =sowberry =holly berry (p?),
  12, 93, 94, 96, 129
bayberry *Myrica cerifera;* =wax myrtle =candleberry =snowberry =waxberry, 103
bearberry *Arctostaphylos uva-ursi;* =fenberry (England) =upland cranberry =arberry
  =Mountain boxberry =hog cranberry =mealberry =Uva-ursi (D), 13, 59, 60,
  64, 66, 134
bear huckleberry =blueberry, 54
berberry =barberry, 94
bilberry *Vaccinium myrtillis;* =English blueberry, 53
blackberry *Rubus villosus, R. occidentalis, R. canadensis;* =dewberry =goutberry
  =cloudberry, etc. (A), 8, 9, 10, 11, 15, 16, 31, 32, 46, 48, 49
black cap =black raspberry, 45
blackhaw *Viburnum prunifolium* or *V. rufidulum;* =nannyberry =sloe, 64

black mulberry *Morus nigra*, 77

black raspberry *Rubus* var.; =thimbleberry =blackcap, etc. see also blackberry (A), 42

blueberry *Vaccinium myrtilloides*; =American huckleberry =whortleberry, etc., 10, 16, 53

bog cranberry *Vaccinium vitis-idaea*; see cranberry

boxberry =wintergreen, 120

boysenberry *Rubus* hybrid; see also blackberry, 49

bramble, brambleberry a generic name for the blackberry and its relations, 31

buckberry =huckleberry, 54

buffalo berry *Elaeangus angustifolia* (D), 133, 135

bulberry =mulberry, 77

bunchberry *Cornus canadensis*; =crackberry =crackerberry (D), 13, 137

candleberry =bayberry, 13, 104

cape gooseberry *Physalis peruviana*, etc.; =tomatillo =jamberry =strawberry tomato =pohaberry =ground cherry, etc. (T), 138

carberry = (English) gooseberry, 70

cedarberry =juniper, 99

checkerberry =checkberry =wintergreen, 120

chokeberry *Aronia arbutifolia*, 140

cloudberry *Rubus chamaemorus*; =knot berry see also blackberry, 46

coralberry =Indian currant, 145

cowberry =cranberry, 60

crackberry =bunchberry, 137

crackerberry =bunchberry, 137

cranberry *Vaccinium macrocarpum*; =cowberry =lowbush cranberry =lingonberry =bog cranberry =squashberry =foxberry, 10, 12, 59, 60, 64, 66

crowberry *Empetrum nigrum*; =curlew berry, 12, 141

curlew berry =crowberry, 12, 141

currant *Ribes rubrum, R. nigrum*, etc.; =skunkberry (B), 9, 73

dangleberry =huckleberry, 52

deerberry =partridgeberry =huckleberry =wintergreen, 11, 13, 53, 54, 121, 144

dewberry see blackberry

dingleberry *Vaccinium erythrocarpum*; see cranberry

dogberry =gooseberry, 69

dwarf cape gooseberry =ground cherry =husk tomato =strawberry tomato (T) see cape gooseberry

elderberry *Sambucus canadensis, S.nigrum, S.racemosa*, (p, D), 85

farkleberry =huckleberry, 53

fenberry see cranberry

foxberry =cranberry =blueberry, 55

goldenberry =cape gooseberry, 138

gooseberry *Ribes grossularia, R. uva-crispa, R. missouriense, R. sanguineum*; =dogberry =carberry =peaberry =thapes =grozer =skunkberry, 13, 69

goutberry =blackberry, 12, 34

groundberry =wintergreen, 120

grouseberry =wintergreen, 121

hackberry *Celtis occidentalis*; =hagberry =bird cherry =sugarberry (D), 10, 11, 142

hagberry =hackberry, 142

haw, hawthorn berry *Crataegus species* (P), 64, 109

highbush blueberry *V. corymbosum*, 52

highbush cranberry *Viburnum trilobum* (D) =cranberry, 64, 66

Himalaya berry see blackberry

hoboys =strawberries, 26

hog cranberry =bearberry, 60, 134

holly berry see barberry

huckleberry *Gaylussacia*, all vars. *Vaccinium myrtillis, V. tenellum, V. frondosum*, etc.; =bilberry
  =whortleberry =sunberry =buckberry =wonderberry, =hurtleberry, etc., 13, 52, 54

hurtleberry =huckleberry, 54

ibimi Indian name for cranberries, 12

Indian berry =wintergreen and many other berries, 10

Indian currant *Symphoricarpos vulgaris;* =coral-berry (B), 9

Indian strawberry *Duchesnea indica;* =mock strawberry =snakeberry, 25

inkberry =poke, 125

jamberry *Physalis ixocarpa;* =tomatillo (T) see cape gooseberry

jaundice berry see barberry

josta berry hybrid of currants and gooseberries, 73

Juneberry see serviceberry

juniper berry *Juniperis virginiana* or *J. communis*, etc.; =cedarberry =red cedar
  =medicine berry, 99

knotberry see cloudberry

lemonade berry see sumac

lingberry see mountain cranberry

lingonberry =European cranberry, 60

loganberry see blackberry

lowbush cranberry see cranberry

mayapple =raccoonberry, 106

mayberry see blackberry

mayhaw see hawthorn

mealberry =bearberry, 134

medicine berry *Juniperus sabina*, also *J. virginiana;* =savine, 100

mooseberry *Viburnum edule;* =squashberry see cranberry

mountain boxberry see bearberry

mountain cranberry *Viburnum opulus, Vaccinium macrocarpon;* =crampbark =squawbush
  =American cranberry, 11, 12, 59, 64, 134

mountain grape see barberry

mulberry *Morus alba, M.nigra, M.rubra* (A), 10, 11, 13, 77

nannyberry *Viburnum lentago;* =sheepberry =wild raisin, 64, 66

necklaceberry =baneberry, 13

nectarberry *Rubus* variety, 49

ollalie berry *Rubus* hybrid, 49

oregon grape see barberry

osoberry *Osmaronia cerasiformis;* =Indian plum (D), 9

palmetto berry *Serenoa serrulata;* =serenoa, 119

partridgeberry *Mitchella repens* or *Gaultheria procumbens;* =deerberry =scarletberry =spiceberry
  =checkerberry =twinberry =stinkberry, 11, 12, 121, 143

peaberry =gooseberry, 70

pigeonberry *Duranta repens;* see also wild cherry (p), 125, 129, 144

pimbina *Amelanchier* var.; also *Viburnum opulus* and other vars., 83

pohaberry *Physalis pruinosa* (T), see cape gooseberry

pokeberry *Phytolacca decandra;* =pigeonberry =poke =garget =inkberry =pokeberry (p), 125

pudding berry =bunchberry, 137

purgeberry *Rhamnus cathartica;* =buckthorn (p), 12

rabbit berry see buffalo berry

raccoonberry *Podophyllum peltatum;* =mayapple =mandrake =yellowberry =wild lemon (B), 106

raspberry *Rubus strigosus,* etc.; =blackberry =black cap =black raspberry =brambleberry =cloudberry =loganberry =red raspberry =salmonberry =thimbleberry =wineberry (A), 39

redberry =buckthorn =wintergreen, 120

red mulberry *Morus rubra,* 79

red raspberry *Rubus idaeus;* see blackberry

rockberry *Empetrum Eamesii;* see crowberry

Rocky Mountain grape see barberry (B)

roeberry =salmonberry see blackberry

roseberry archaic name for rose hips, 113

rosehips *Rosa,* var. (B), 113

roxberry =wintergreen, 120

salmonberry *Rubus parviflorus;* =thimbleberry, 34, 48

sarvis, sarvis-berry =serviceberry, 81

saskatoon *Amelanchier alnifolia;* see serviceberry

savine (also sp. **savin**) see juniper berry

saw palmetto berry *Serenoa serrulata;* =saw palmetto, =serenoa (D), 119

scarletberry =partridgeberry =bittersweet (*Solanum dulcamara*), 143

serviceberry *Amelanchier canadensis,* etc.; =shadbush =shadberry =Juneberry =service apple =wild raisin =pimbina =saskatoon (P), 9, 10, 81, 82

shadberry see serviceberry

shadblow see serviceberry

shadbush see serviceberry

sheepberry =nannyberry, 64, 66

silverberry *Elaeagnus commutata,* 136

skokeberry =poke, 125

skunkberry =gooseberry =currant, 69, 73

sloe, sloeberry =black haw =wild plum, 66

snakeberry =baneberry =Indian strawberry, 25

snowberry *Symphoricarpus albus, Chiogenes hispidula;* =bayberry =creeping snowberry, 145, 146

southern dewberry see blackberry

southern mountain cranberry see cranberry

sowberry =barberry, 93

sparkleberry =huckleberry, 53

spiceberry *Myrtilis,* var., 103

squashberry =cranberry also *Viburnum edule;* =mooseberry, 60, 66

squawberry see huckleberry =deerberry see also partridge berry, 54, 144

squaw huckleberry =huckleberry, =partridgeberry, 53, 54

stinkberry =partridge berry (*M. repens*), 144

strawberry *Fragaria,* var.; 9, 10, 11, 23

strawberry raspberry *Rubus illecebrosus;* =balloonberry, 46

strawberry tomato see cape gooseberry

sugarberry =hackberry =bird cherry =hagberry, 142

sugarpear see serviceberry

sumac (also sp. **sumach**) *Rhus typhina, R. glabra* (D); 12, 13, 117

sunberry =huckleberry, 52

swampberry =huckleberry, 69

tayberry *Rubus* hybrid, 49

teaberry *Gaultheria procumbens;* =boxberry =checkerberry =deerberry =grouseberry
   =roxberry =pipsissewa see also partridge berry =wintergreen, 11, 12, 13, 120

thimbleberry *Rubus occidentalis;* see blackberry

twinberry =lonicera, var. (p?) =partridge berry, 129, 144

vinegar berry =sumac, 13, 117

waxberry =bayberry, 103, 104

western thimbleberry *Rubus nutkanus;* see blackberry

white blackberry see blackberry

white mulberry *Morus alba,* 77

whortleberry see huckleberry

windberry see cranberry

wineberry *Rubus phoenicolasius;* =Japanese raspberry =whortleberry =currant =gooseberry
   see also blackberry, 48

wolfberry *Symphoricarpos occidentalis,* 146

wonderberry =huckleberry, 52

wood strawberry *Fragaria vesca;* see strawberry

yellowberry =mayapple, 106

youngberry *Rubus* hybrid, 49